HOW MICROPROCESSORS WORK

By Steven Kaminsky

Table of Contents

CHAPTER ONE
WHY MICROPROCESSORS?

Arguably the most important invention of the modern world is the microprocessor. Arm claims to have sold 30 billion of their microprocessors. Microchip sells 1 billion of their microprocessors per year. Yes, that's "billion" with a "b." Where in the world are all these things?

The numbers tell us they must be in many more places than just our computers. Microprocessors run smart phones, networks, web sites, cars, laptops, appliances, factories, etc. We are often unaware that, at the heart of devices we use every day, a hidden, or "embedded," microprocessor controls the device. It might be in a toy or in a coffeemaker. Your car may have 12, or 50, or 100 microprocessors. For example, one may be dedicated to collecting the air pressure in each tire.

And yet, even though we use all these devices, we don't understand how they work. They seem like magic. That's because we don't understand how microprocessors work. We don't understand digital electronics.

It's time to stop being ignorant. This book purports to lift the veil of secrecy, the clouds of ignorance hiding this wonderful device. I have every confidence that, when you learn how these microprocessors work, you will find them to be even more wonderful than before. When you learn the secret of how a magic trick works, it often becomes mundane, as in, "is that all there is to it?" When you learn how microprocessors work, I believe you will be even more in awe of them.

The first confusion we need to address is the distinction between the terms "microprocessor" and "computer." When we talk about a computer, we usually mean a device designed in such a way that it can do multiple tasks for us. We can change that task by simply giving the device a new set of instructions: a new program.

Our early computers were large devices. They were built from vacuum tubes and from many discrete components. These were built into sub-circuits, and the sub-circuits into circuits, and the circuits into circuit boards, and the many circuit boards worked together to function as the computer described in the previous paragraph.

In the late 1940's tiny transistors were invented, as replacements for vacuum tubes. Transistors were first made of germanium, then later of silicon. Then, engineers thought, why not create more than one transistor on a wafer of silicon? Why not connect the transistors and other components on the wafer of silicon?

Around 1960, they succeeded, calling this device an "integrated circuit," because an entire functioning electronic circuit could be fabricated, and integrated into one small device. Called a "chip" or "IC," an integrated circuit was a piece of silicon enclosed in plastic, with legs or "pins" that connected the internal circuitry to the outside world. What you could hold in the cup of your hand or balance on the tip of your finger replaced entire circuit boards full of many components.

At the time, many people knew what was coming next: putting the "brains" of a computer, the central processing unit (CPU), into one integrated circuit. Steve Wozniak couldn't wait. Why, if he could get one of these CPU's, he could add some memory and input/output devices to it. He could build a home computer around it, and call it the Apple Computer.

They called this new CPU on a chip a "microprocessor." Companies like Intel and Motorola made the microprocessors that brought this dream to fruition. So, a microprocessor is the brains, the essence, the CPU (central processing unit) of a computer, made really small. A microprocessor is the heart, the computing part of a computer.

I had difficulty choosing a name for this book. Was this going to be a book about how computers work, or about how microprocessors work? I realized it should be about how microprocessors work, because computers are now a subset of all microprocessor-based devices.

Here's why we have more microprocessors than computers. Historically, computers came first. Computers--digital electronic multipurpose devices that can run different programs--were large devices. Next, engineers put the processing part of the computer into a small integrated circuit, and called it the microprocessor. We now could have a personal, home computer. At this point in history--and from here on--we all know what we mean when we say the word, computer.

But next, something strange happened. Engineers started using--and still use--this same microprocessor for other designs that do not fit our definition of a computer. They use it in designs that don't even require a microprocessor: that could be designed using solely analog or digital electronics components and techniques. The fuel injection systems of cars existed for years using mechanical timing, before anyone invented microprocessors. Now, fuel injection systems are microprocessor-based.

New generations of microprocessors were designed that were more than just the CPU part of the computer. They included the memory and input/output parts of a computer, in addition to the CPU. Now, these newer microprocessors were more like a "computer on a chip" than a "CPU on a chip." Sometimes, we call these "computers on a chip" *microcontrollers*, while calling "CPU's on a chip" *microprocessors.*

If you told an engineer from the 1950's that one day a clock or a toaster oven design would be implemented using a "computer on a chip," they would have laughed. To them, it would have been like using a sledge hammer to pound a tack in the wall.

So, this creates some confusion in terminology. We have gotten to the point where engineers are using a "computer"--a microcontroller-- to create things that are not computers! We know what a computer is, and we know what a clock is, and we know that a clock is not a computer. And yet, we find out that, today, a clock has a "computer" in it. We find that everyday things that are not computers have little computers-like devices--microcontrollers--controlling them.

In short, computers became microprocessors, then microprocessors became--almost everything. I think the best way to avoid confusion is to define a computer as a digital electronic multipurpose device that can run different programs. Single-purpose devices running one program we can call microprocessor-based systems. Remember, though, that a computer, too, is a microprocessor-based system. It just runs more programs, more tasks. Do you see now why it is so important to understand microprocessors? They are everywhere.

Why did engineers start using microprocessors to design things that aren't computers? It just became easier. If you spent years learning to play the guitar well, do you really want to learn to play the piano in order to play your song? If you know how to design with microprocessors well, do you really want to design your next project with a unique analog or digital design? A unique, non-microprocessor design might be more efficient, and the microprocessor might be overkill. Still, you know how to design with the microprocessor quickly, it flexibly adapts to almost any project, it efficiently replaces a lot of circuitry with software, components are available and compatible with each other, and the cost and power consumption has fallen so much that overkill is no longer an issue.

This book is aimed for the average, non-technical, person. However, I have found that even computer programmers do not really know how the devices they write their programs for, really work. In fact, they don't need to know. You may have heard of the term, "higher level languages," in reference to programming languages like C++ and Java. This term actually means that the language is a step above machine's language, so that people can program the machine without having to understand the machine. The microprocessor is the machine.

So, this book is for the programmer, too. And, it's for the IT people, and the networking people. I have found that the people with the deepest understanding of programming, IT, and networking are people who come from a background of electronics. They see the big picture, from the ground up. The microprocessor and digital electronics are the ground, the foundation.

CHAPTER TWO
BINARY

Look at the word, "microprocessor." "Micro," of course, means small, really small, as in microscopic. What is being "processed," though? The answer is: information, as in data or numbers. The information is represented in binary format.

I am sure you have heard that everything in modern, digital computers is in binary: ones and zeroes. You should wonder, what are these ones and zeroes? Are they physical things? Why ones and zeroes?

Let's forget about microprocessors for a moment. "Binary" means a system of two states. A two-state system is absolutely the simplest information system imaginable. A one-state system contains no information; it is undifferentiated.

A two-state system seems, at first glance, barely better than a one-state system. But, looking deeper, we see the variety of information we can represent or encode in a two state system. The two states can stand for: one or zero, true or false, high or low, odd or even, go or stop, left or right, light or dark, black or white, good or evil, yin or yang, and so on. We can control a simple machine like a motor with a binary, two-state system: motor on or motor off.

Looking even deeper, the second thing we see is that this system grows enormously in its capability when we expand it. In the above paragraph, I described a simple two-state system of one item. We call this one item a "bit." One bit can be in two states. Defining the information in terms of black and white, a one bit system can hold two state conditions: black or white.

However, if we expand the system by grouping two of these bits together, we double the amount of information our system holds. A two-bit system holds four state conditions: black-black, or black-white, or white-black, or white-white.

If we expand again by grouping three of these bits together, we quadruple the amount of information our one bit system holds. A three-bit system holds eight state conditions: black-black-black, or black-black-white, or black-white-black, or black-white-white, or white-black-black, or white-black-white, or white-white-black, or white-white-white.

If you want to calculate the number of unique pieces of information that a binary system is able to represent, count the number of bits you have grouped together, then take 2 to that power. For example, you probably have heard of the term, "byte." A byte is a grouping together of eight binary bits. If you take 2 to the 8th power, written as 2^8, you get 2x2x2x2x2x2x2x2 which equals 256. You now have a system that can represent 256 unique informational states. Any one of these states is different from the other 255 states. Black-black-black-black-black-black-black-black is different from black-white-black-white-black-white-black-white.

The growth in power over the years of personal computers is reflected in the constant growth of the number of bits grouped together. The earliest Intel microprocessors were 4-bit systems, then 8-bit, then 16-bit, then 32-bit, then 64-bit. One group of 64 bits can uniquely represent 2^{64} or 18,446,744,073,709,551,616 informational states. One of these states might look like: black-black.

So, a binary system starts out being ridiculously simple: a 1-bit system capable of only 2 states. Yet, if we clump together just 64 of these bits, we now have a ridiculously powerful group capable of 18,446,744,073,709,551,616 different states!

One can implement a one-bit, two-state binary system in many ways. For example, a coin is a one-bit, two-state binary system. The two states are: heads or tails. Another is a light: on or off. A match is lit or not lit. A true-false question is binary. A piano key is pressed or not.

The question becomes, what is the best way to implement a one-bit, two-state binary system? In the history of computers, several implementations have been used. In early mainframe computers, individual memory bits were ring magnets. Each one could be magnetized either clockwise or counter-clockwise (binary) by forcing current forwards or backwards through a wire coiled around the ring.

Hard drives are also magnetic storage media. Their disks are made of ferromagnetic material. A predefined spot on a disk acts as binary bit. The spot is magnetized in the forward (north pole-to-south) or reverse (south pole-to-north) direction by an electromagnet moved by a motor. This same spot can be read by a wire coil as the spot moves past, inducing either forward or reverse current flow.

Compact discs and DVD discs use light to implement a one-bit, two-state binary system. A predetermined spot on the disc acts as a binary bit. A laser light is used to change or not change the reflective properties of the spot. The laser is also involved in retrieving the binary state of that spot. When the reading laser light is aimed at the spot, it reflects differently, depending on whether the writing laser had previously modified it or not.

So, what is the best way to implement a one-bit, two-state binary system? The answer is a surprising one. The answer is, "PRESSURE." Of course, we are talking about a special kind of pressure.

Let's pause for a moment and think of how we could communicate using pressure. Here is a thought experiment. If you were to lie on your stomach and close your eyes, how could people communicate with you in binary, using only pressure?

Here is how they could do it. Designate different spots on your body, say the back of each knee, and the back of each elbow, as pressure points, acting as binary bits. Then, four people could communicate with you by either pressing or not pressing those four designated spots. 2^4 or 16 different messages could be communicated to you in this way. For example, one distinct message could be: left elbow pressed, right elbow not pressed, right knee pressed, left knee not pressed.

What would these 16 messages stand for? Whatever you choose. In the morning they could stand for what breakfast will be served. In the evening they could stand for how many runs the home baseball team scored. Binary systems, including computers, use grouping of bits at different times, in different subsystems, for different purposes.

Of course this is not the kind of pressure that microprocessors use. Let's get a little closer to that by imagining a binary system encoded by something we are familiar with: water pressure. Water pressure comes from a distant water pump, with a propeller inside that pushes on the water. The pressure also may come from gravity, via a water tower or reservoir higher than your house. Let's look at how a water faucet works.

When the control valve is turned off, no water flows, but there is pressure at the faucet. The water is pressing on the closed valve.

When the control valve is turned on, water flows, but there is no pressure at the faucet. The valve is out of the way, so there is no water pressing on the valve.

So, in our one-bit binary system, the water valve, we have two states: water flowing, or water not flowing. But, we could also look at it as two states: pressure off or pressure on. In fact, the states are inverted or opposite from each other: flow is off when valve pressure is on, and flow is on when valve pressure is off. Figure 2-1, below, summarizes this concept.

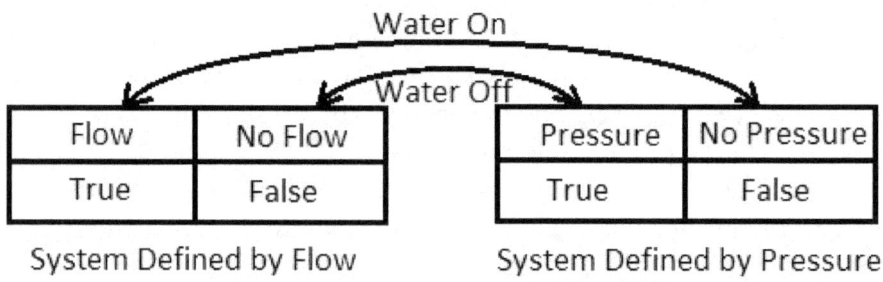

FIGURE 2-1

In our microprocessor-based system--in fact in all digital electronics--we will find that the meaning is encoded in the pressure, rather than in the flow. But, that is getting ahead of ourselves.

Imagine now a four-bit binary communication system based on water pressure. We put four water faucets in a row. Any one faucet can be independently turned off or on. Just as with the pressure-on-the-knees-and-elbows system, we can encode 2^4 or 16 different messages.

Let's look now at a subtle, but important difference. Name the valves, from left to right, A, B, C, and D. Turn valves A and C on, so that water flows. Turn valves B and D off, so no water flows. If we choose to define each bit as True or False, we instinctively would assign water flowing as True, and water not flowing as False. So, in the above scenario, we would describe A-B-C-D as True-False-True-

False, as in Figure 2-2, part (a), below:

A	B	C	D
Flow	No Flow	Flow	No Flow
True	False	True	False

4-Bit System Defined by Flow

(a)

A	B	C	D
No Press-ure	Press-ure	No Press-ure	Press-ure
False	True	False	True

Same 4-Bit System, Defined by Pressure

(b)

FIGURE 2-2

This would intuitively be our choice, because we can see water flowing. If we wanted to, we could have defined our system in terms of valve pressure, instead of water flow. That would be a less intuitive choice, because we can't see valve pressure. But, if we had chosen to define valve pressure on as True, and valve pressure off as False, then in our above scenario we would describe A-B-C-D as False-True-False-True, instead of True-False-True-False. Figure 2-2, part (b) displays this different way of defining the same system, in the same state.

Again, in our microprocessor-based systems, we will find that the meaning is encoded and defined by the pressure. In the microprocessor case, the flow is not the more obvious choice, because we can't see the flow. On the other hand, we can't see the pressure either, with our eyes. However, we can measure both the pressure and flow. In fact, in electronics, the pressure is easier to measure than the flow.

I have gone through these examples so that you can get more comfortable with the idea of storing binary information as pressure. I chose these examples because they are forms of pressure we can see and feel. We must now move beyond the familiar to the unfamiliar world of the atomic, or rather, the sub-atomic: to the world of the electron. For, the kind of pressure used to store binary information in microprocessors is: ELECTRIC PRESSURE.

CHAPTER THREE
VOLTAGE AND CURRENT

In digital electronics, the name of the device that is either in the "electric pressure on" or "electric pressure off state" is the *transistor*. The transistor is similar to the faucet, which is either in the "water pressure on" or "water pressure off" state. Like the faucet, which allows water to flow, the transistor allows electric current to flow. Like the faucet--in which water flows when pressure is gone at the valve-- the transistor allows electric current to flow when the electric pressure is gone at its valve. Like the faucet--in which water stops flowing when pressure is present at the valve--the transistor stops the flow of electric current when electric pressure is present at its valve. In fact, the British called the precursor to the transistor a "valve." Americans called it a "tube."

The previous paragraph merely describes how the transistor is analogous to a water faucet. But, it used terms that require explanations. You need to know what "electric pressure" is. You need to know what electric current is. You need to know how a transistor acts as a valve, or more precisely, as a switch. So, the next section of this book acts like a basic electricity and electronics course.

The previous examples of binary information devices have some serious limitations. Obviously, they are limited by their size. Faucets are too big. Coins can't communicate with other coins. The devices are also limited by their flexibility. Bits on CD ROMs are tiny, but they can't communicate with each other.

We need a system in which information can be moved quickly from one part of the system to another part. We need a system in which the state of one part of the system can quickly and efficiently affect and change another part of the system. We need a system in which events can happen sequentially. We need a system that runs and changes on its own, rather than having every bit controlled by a human flipping a switch--like the telegraph--or a human turning on a faucet. Add to this list the need for the system to be low-powered and very compact.

These needs can all be met by a binary system based on electricity.

Changes in electrical pressure happen very quickly. With the invention of the transistor, then the integrated circuit, digital electronics systems can be very small.

So, the first thing we need to do is explain electrical pressure. Water pressure comes from a water pump pushing on the water. Electrical pressure is a bit harder to comprehend. What is it about electricity that can "push?" What gets pushed?

We can see the effects of magnetism. We have all probably played with magnets as children. We can take two magnets, and make one of them move across a table to the other. The one magnet moves the other. We can turn one magnet around, and then try to force the magnets together. We find that the magnets very strongly push away from each other. We have firsthand knowledge that a magnetic force field can push or pull other magnets or ferromagnetic material.

It turns out that there is another force field: the electric force field. It is very similar to the magnetic force field. It, too, can make two pieces of metal move towards, or away from, each other. It makes a balloon rubbed against our sweater stick to us. It makes those nasty, white, "peanut" packing materials stick to everything. It makes the electron beam move across our old CRT TV's and computer monitors. It makes lightning move through the air. It is a fundamental force of nature.

Voltage is the name we give to this electric pressure, this electric force field. Sometimes we call it electromotive force, or EMF, for short. "Electromotive force" is a perfect name, because voltage is a "force that moves electrons." It tells us that voltage is a force; that the force can move things; and that electrons are the things that get moved. We should say "voltage on" or "voltage off" instead of "electric pressure on" or "electric pressure off."

In fact, using the term "electric pressure" is a bit inaccurate, because pressure means <u>force</u> per unit area, whereas voltage means energy per coulomb of charge, or work done per coulomb of charge. Since work is force times distance moved, then *voltage = <u>force</u> x distance / coulomb*. This equation is the mathematical representation of calling voltage an <u>electromotive force</u>. Coulombs is the number of electrons moved (<u>electro</u>), and distance is the measurement of movement (<u>motive</u>). So, pressure and voltage have in common a "force" at the heart of their equations. I use the word "pressure" because it's something we all can feel, like when we press on something and see it move.

To understand voltage--and current for that matter--we must understand the atom. I use the term, "understand," loosely here. I don't feel that we can truly understand the atom, or the electron, or light, or gravity, or electromagnetic waves, or any of the fundamental elements of physics. Our brains best understand the things of the middle range of the universe. We cannot comprehend the incredibly small, like atoms, or the incredible big, like the universe. There is no reason for us to be designed to comprehend these things.

So, we usually use models, based on things we can understand, to help us explain things we can't understand, like atoms. The simple Bohr model of the atom best helps us to understand how voltage comes from atoms.

The Bohr model represents the atom as like a planetary system. In a planetary system, there is a large, massive sun at the center. Orbiting around the sun are the planets, smaller and less massive than the sun. Another force, gravity, pulls the planets toward the sun, keeping them from flying away into space. (Ignoring Einstein and quantum, for simplicity.)

Per this model, the electrons are like the planets. The protons, at the heart of the atom, are like the sun. Electrons are less massive than the protons. The weak nuclear force keeps the electrons pulled toward the protons at center, or nucleus, of the atom.

Unlike gravity, but more like magnetism, we say that electrons have a negative charge (-). Protons have the opposite charge: positive (+). "Like" charges repel, and "unlike" charges attract. This means that there is a fundamental force of nature such that every electron (-) pushes all other electrons (-) away from itself. Every proton (+) pushes all other protons (+) away from itself. This is what we mean when we say, "like" charges repel. Every electron (-) and every proton (+) pull toward each other. This is what we mean when we say, "unlike" charges attract.

So, the electrons on the outside of the atom try to stay away from each other, as much as possible, while at the same time they are attracted toward the protons in the center of the atom. If you think about this for a while, you should wonder why the protons, crammed together in the nucleus and repelled by each other, don't fly apart. Since this doesn't make sense, we say that they are held together by a force of attraction, stronger than the force of repulsion. We call it the "strong nuclear force." Sounds suspiciously convenient.

The strength of the force in one electron or in one proton is very small, but very real, nonetheless. And, there are so many electrons, so many protons. And, their strength is cumulative. Witness the power of a lightning strike. If we add a lot of extra electrons to one piece of metal (so that it is excessively negative), and remove electrons from another piece of metal (so that it is excessively positive), those two pieces of metal will move towards each other, just like two magnets.

One atom may have many protons and electrons. And within this atom, many pressures exist. But the pressures are very small. And, the net pressure is zero, since in one atom there are the same number of electrons as there are protons, and the positive charge of one proton exactly equals the negative charge of one proton. Putting a whole bunch of atoms together doesn't change things, because their net pressure is still zero: a bunch of zero net pressure atoms adds up to zero net pressure.

How do we create the electric pressure we've been talking about? We have to un-balance the atom. One of the definitions of voltage (EMF) is "a separation of charge." If we remove an electron from an atom--separate the charge--the remaining atom no longer has exactly the same number of electrons as protons. It is not in balance. It no longer has zero net charge. It has one more proton than electrons, so it has a net charge of +1, or +1 times the force of one proton.

One the other hand if we add an electron to an atom--separate the charge from elsewhere--the atom no longer has exactly the same number of electrons as protons. It is not in balance. It no longer has zero net charge. It has one more electron than protons, so it has a net charge of -1, or -1 times the force of one electron.

In each of the above cases, the atom has net pressure, either positive or negative. We call this atom a positive ion or a negative ion. The more ions we put together, the more voltage (pressure) we have. We have found what we have been looking for: voltage.

The question now changes to: how do we separate charge, or equivalently, how do we create voltage? One way to do it is through brute force. We can use our muscles and rub things together. If you rub a balloon on your wool sweater, the balloon may then be able to stick to a wall. You have forced a separation of charge. Because of their atomic structure, wool is very willing to give up electrons and balloons are very willing to gain electrons.

The balloon does truly have voltage. It is called "electrostatic" voltage, or "static electricity". It's the same electricity that causes the static shock we feel when we touch a door knob. But this voltage is temporary. When the charged item touches something that has the opposite charge, the light-weight electrons move to the positive object: the static voltage is gone. You have to replenish the charge by rubbing, again.

We would like a self-replenishing source of voltage. We call this self-replenishing voltage source a battery. See the picture below. A typical battery has two rods immersed in a fluid called an electrolyte. The rods are made of dissimilar metals, like nickel and cadmium. A chemical reaction occurs, causing the electrons from one metal rod to move through the electrolyte to the other metal rod. The rod that gains electrons is called the negative terminal or cathode. The rod that loses electrons is called the positive terminal or anode. In a ni-cad battery, the anode is made of nickel, the cathode of cadmium, and the electrolyte of potassium hydroxide.

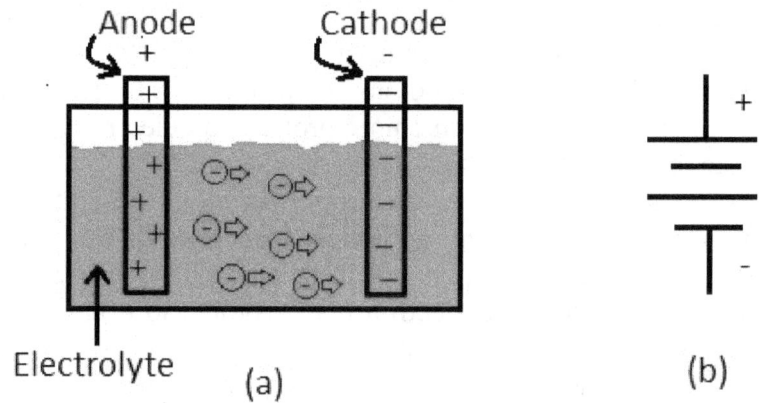

FIGURE 3-1

Figure 3-1, part (a), shows the process of charge separation when the electrolyte is poured into the battery. Figure 3-1, part (b), shows the electrical symbol for a battery.

The battery now has voltage: pressure to move something. What does voltage move? Voltage moves electrons. This movement of electrons is called, *current*. Current is electrons moving. Not moving within an atom, but moving out of an atom and going elsewhere. But to where do they move?

To answer this question, we have to pause and look at the world of chemistry, atomic structure, and the periodic table. The periodic table describes atoms. Each atom is numbered, in ascending order. The numbers stand for how many electrons, protons, and neutrons each atom has. Hydrogen is numbered "1" because it has one electron, one proton, and one neutron. Oxygen is numbered "8" because it has eight electrons, eight protons, and eight neutrons. This tells us that what makes one element different from another is the number of electrons, protons, and neutrons in its atom.

The columns in the periodic table have significance, especially to the electrical properties of matter. The electrons in a specific atom arrange themselves in groups, from inner groups--or shells--to middle groups, to outer groups of electrons. A column in the periodic table describes elements made of atoms whose outer shells exhibit similar properties.

The outer--or valence--shell of a specific atom will have between one and eight electrons. For example, silicon atoms have four electrons in their outer shell. Copper atoms have one electron in their outer shell.

Let's go back to the question of where electrons can move: where current can flow. We find that, electrically, there are three categories of materials, conductors, semiconductors, and insulators.

Conductors are materials through which current can very easily flow. A conductor has one electron in its outer shell. The one electron never attains a symmetrical balance of attraction-repulsion within the atom, so it is very free to move away from the atom. In the periodic table, you will find conductors grouped together in the eleventh column: copper, silver, and gold.

Semiconductors are materials are materials through which current can somewhat flow. A semiconductor has four electrons in its outer shell. These four electrons have a moderate symmetrical balance of attraction-repulsion within the atom, so they can be moved away from the atom if adequate outside pressure is applied. In the periodic table, you will find semiconductors grouped together in the fourteenth column: carbon, silicon, and germanium.

Insulators are materials in which current does not flow. An insulator has eight electrons in its outer shell. These eight electrons are in a strong symmetrical balance of attraction-repulsion within the atom, so they are not free to move away from the atom. In the periodic table, you will find insulators grouped together in the eighteenth column:

helium, neon, and argon. Also, when elements combine together, they tend to share outer shell electrons and align in such a way that each element's outer shell has eight electrons, becoming an insulator. That's why sodium chloride (table salt) is a good insulator. So are plastic, glass, rubber, and air.

Returning now to electrical current, let's imagine our battery, with extra electrons (negative ion atoms) congregating on the negative terminal, and missing electrons (positive ion atoms) on the positive terminal. The electrons on the negative terminal would really like to get away from each other. They would like to get to the positive terminal, so that both terminals would be neutral. The atoms could return to their natural, neutral, non-ionic state. But they can't go through the battery. It was the chemical reaction in the battery that did all the work in creating this imbalance in the first place, storing the work as potential electrical energy. You may have heard of this referred to as electrical potential.

The electrons could escape from the negative terminal if we put a conductor, like a copper wire, between the positive and negative terminals of the battery. See the picture below. (WARNING: Don't do this, it could blow up the battery. I will explain about the need for electrical resistance later.) Remember the previous discussion about conductors like copper? Their atoms have one very loosely bound electron in their outer shell. They are very easily moved by electrical force. The instant you connect the wire to the battery terminals, these electrons feel the attraction of the positive ions in the positive terminal and they feel the repulsion from the negative ions in the negative terminal. They move! They move towards the positive terminal.

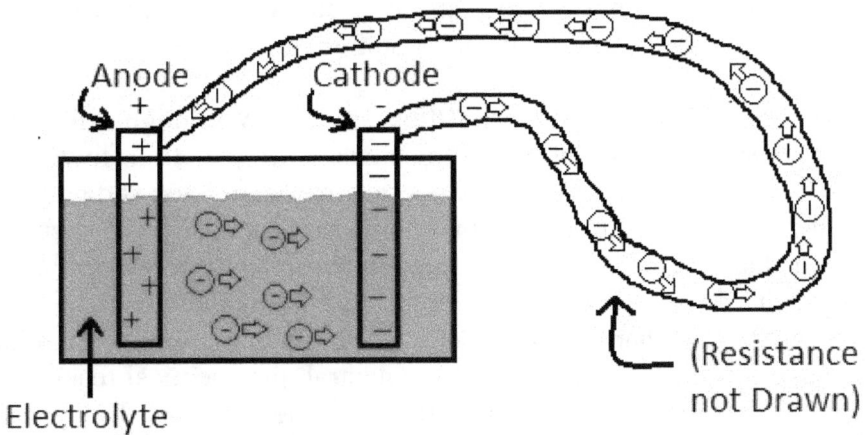

FIGURE 3-2

Let's stop to enjoy, appreciate, and wonder at this. We call this current, because it is like water current in a stream. But this is movement in a wire. The very wire has motion--at the atomic level-- inside it. The solid wire has electrons flowing inside it. We have answered our question, "Where does current flow?"

We have devices called ammeters that measure the amount of movement inside a wire. They measure the number of electrons that move through a cross-sectional area of wire, per second. The unit of measure is the "ampere." An ammeter might measure one ampere, or one amp, of current in a wire coming into this computer. One amp is 6.25×10^{18} electrons per second passing a cross-sectional area of that wire. That's 6,250,000,000,000,000,000 electrons passing by, every second! (We call 6.25×10^{18} electrons a *coulomb*. So, one ampere is one coulomb per second.)

So, the electrons in the atoms that constitute the wire move toward the positive terminal. But the electrons from the negative terminal of the battery are right behind them. Remember, they are repulsed by each other, and were waiting for their chance to escape.

Next, the battery terminal's atoms start going neutral. The negative battery terminal was negative because its atoms had extra electrons. These electrons are leaving, so the atoms become neutral. The positive battery terminal was positive because its atoms were missing electrons. Electrons are now moving in and joining the atoms, so the atoms become neutral.

So, it would seem that current flow is done. You might think that we have a situation similar to electrostatics, where electron flow stops, because the sources of repulsion/attraction are neutralized and gone.

But no, something interesting happens. The battery's chemical reaction that had been finished doing its work in separating charge, starts again. As soon as terminal atoms start going neutral, the battery starts separating charge again. As an atom on the positive terminal acquires a missing electron to become neutral, the chemical reaction in the battery takes it away again. It sends the electron through the electrolyte fluid, and forces it back into an atom on the negative terminal that had just gone neutral because it lost its extra electron.

So, the chemical reaction in the battery restores the separation of charge on the battery terminals (the voltage), during current flow. The electron current flow that moves through the wire continues to move through the electrolyte fluid! So, the circuit is truly a circuit, since electrons even circulate through the battery. Current flow does not stop.

Remember, I said you must not put a wire across the terminals of the battery. Too many electrons would flow. The equation for the amount of current flow is $I=V/R$, where I is current, V is voltage, and R is resistance. Resistance is a measure of the ability of a material to impede current flow. Well, the resistance of a wire is near zero. If you remember a little of high school algebra, you will know that, as the denominator approaches zero, the quotient approaches infinity. Nearly infinite current flows through the wire, or near-infinity times 6.25×10^{18} electrons per second. Something is destroyed: either the battery or the wire.

What you should really put across the battery is a semiconductor. A more manageable amount of current will flow. Carbon is a good semiconductive material. We make components called resistors out of carbon. Rather than a wire, a resistor should be put across the battery terminals to make a moderate amount of current flow.

Now you can see why this subject is called "electronics." The source of voltage is electrons: electrons separated, then packed together. The current, too, is electrons: electrons in motion. Electronics is electrons moving electrons. It's all about electrons.

CHAPTER FOUR
THE FAUCET ANALOGY

I felt that the above, rather lengthy, section on voltage and current was necessary so that you could understand the binary encoding of on-or-off electrical pressure. Next, I would like you to recall that this voltage pressure system is analogous to a water pressure system. You encounter water pressure systems every time you use a water faucet. I would like to use your familiarity with water pressure to help you understand a microprocessor-based system.

Before I introduce the concept of a transistor-as-switch, let's first look at a faucet-as-switch. Look at the picture below.

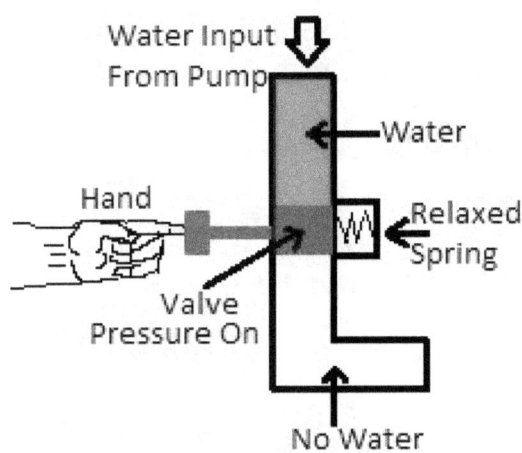

FIGURE 4-1

The above picture is a diagram of an on-off style water faucet, like a drinking fountain. If you don't press the button, no water flows. If you press the button, water flows. It is either on or off. A transistor circuit can be designed to work like this: on or off. This is digital design. "Digital" is a poor term for a binary system.

This is not the kind of faucet you have in your sink. Your sink's faucet is an analog system, because it can be in an infinite number of states. The water flow can be a little "on", or a little more "on", or a little more "on" than that, etc. A transistor can be designed as an analog circuit, too. We would then call it a transistor amplifier. This is not how transistors are used in microprocessors, or in any digital circuits. In digital circuits, transistors are used as on-off switches. Digital circuits have, not an infinite number of states, but just two states: on or off.

The above figure shows the faucet button <u>not</u> being pressed. A water pump, not shown, is applying pressure to the water, forcing it down to the valve. The water cannot get past the valve, because the valve completely blocks it. No water flows from the bottom of the figure. You cannot get a drink.

We would say that the above faucet is *off.* No water is flowing. But we could also say that, *at the valve, pressure* is *on.* The valve feels the full pressure from the water pump. Remember this distinction, because it applies to transistors.

Look now at the figure below.

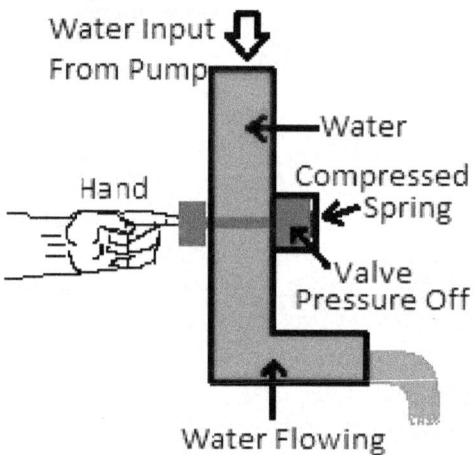

Water Faucet On

FIGURE 4-2

This new diagram shows the faucet button being depressed. The valve moves right, and out of the way. It no longer blocks the water. The water flows out of the pipe, on the bottom of the diagram. You can get a drink.

We would say the faucet is *on*. Water is flowing. But we could also say that, *at the valve, pressure* is *off*. The valve feels no pressure from the water pump, because it is out of the way of the flowing water. If you get confused by transistors, just come back to these basic concepts.

The spring in the above design is a practical consideration. When you stop pressing the button, the spring pushes the valve back to the blocking position, stopping water flow.

So, a transistor is like an on/off water faucet. But, one water faucet or one transistor is just one binary bit, so it cannot encode much information. We can put a group of water faucets next to each other, or group together transistors, and increase the number of unique pieces of information that we can encode. See the figure below:

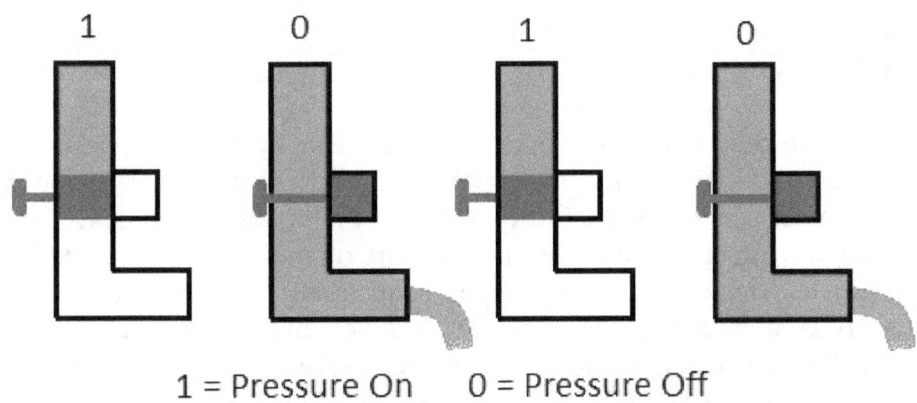

1 = Pressure On   0 = Pressure Off

FIGURE 4-3

In the above diagram, using four water faucets, we have created a four bit binary system. We choose to name our two states "1" and "0". We choose to define the "1" state as the condition in which the valve receives pressure. We choose to define the "0" state as the condition in which the valve receives no pressure. We refer to a four bit binary system as a *nibble* (no joke). The above figure shows the system in the 1-0-1-0 state. This is one of the 2^4 or 16 unique states we can encode. The 16 states are:

0-0-0-0
0-0-0-1
0-0-1-0
0-0-1-1
0-1-0-0
0-1-0-1
0-1-1-0
0-1-1-1
1-0-0-0
1-0-0-1
1-0-1-0
1-0-1-1
1-1-0-0
1-1-0-1
1-1-1-0
1-1-1-1

But we need to do more than this. The above system still requires a hand to press every water faucet. Recall above that we said: "We need a system in which the state of one part of the system can quickly and efficiently affect and change another part of the system. We need a system in which events can happen sequentially. We need a system that runs and changes on its own, rather than having every bit controlled by a human flipping a switch." I would like to take you through a design of such a system using water faucets. Understanding this system should help you to understand the transistor-based system.

Look at the picture below:

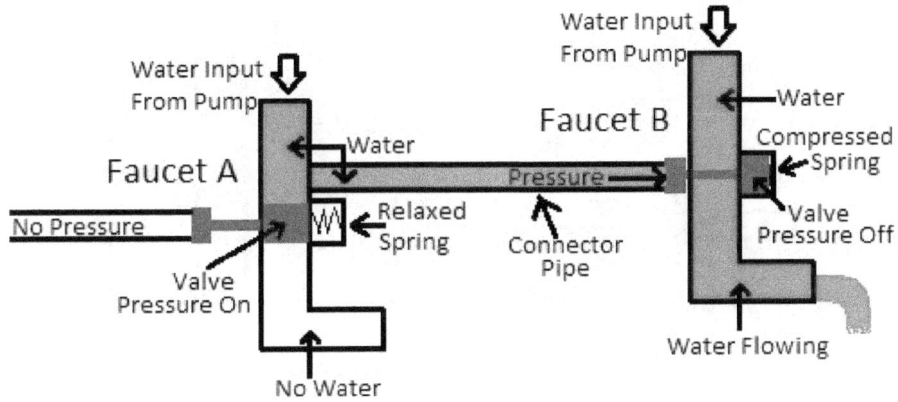

FIGURE 4-4

In this system, the state of faucet B is controlled by faucet A. No human presses the button of faucet B. Nevertheless, faucet B's button is depressed. We have added a connector pipe from faucet A to faucet B. The pressurized water flows down faucet A, but gets blocked by faucet A's closed valve. The water flows from left to right through the connector valve. Since the water is pressurized by the pump, it pushes on faucet B's button. Faucet B's valve opens, its valve has no pressure on it, water gets past its valve, and water flows out of faucet B.

In short, when no water flows from faucet A, water flows from faucet B. Phrased differently, when pressure is on faucet A's valve, pressure is off faucet B's valve.

Now, look at the next picture:

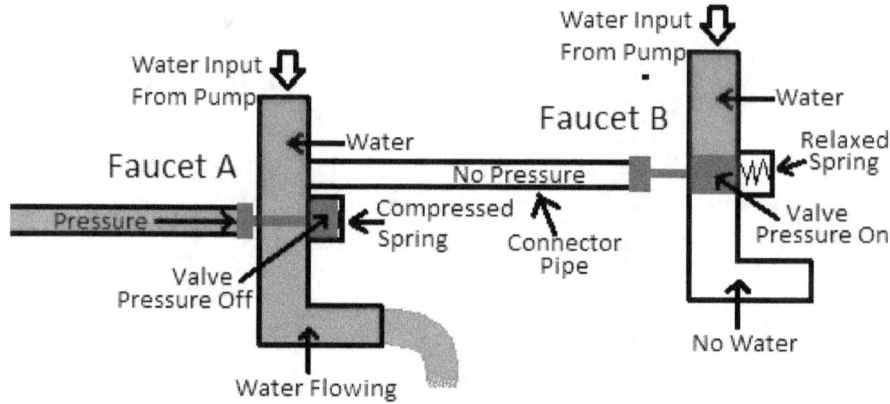

FIGURE 4-5

This system is the same one shown above in Figure 4-4. However, this time, water pressure on the far left enters the system and depresses the button on faucet A. This causes everything in the system to reverse. Now, water flows from faucet A. Now, no water flows from faucet B. Now, no pressure is on faucet A's valve, while pressure is on faucet B's valve.

You are now seeing our need fulfilled: "a system that runs and changes on its own, rather than having every bit controlled by a human flipping a switch." Faucet B is being controlled, not by a human, but by faucet A. If faucet A flows, faucet B doesn't. If faucet A doesn't flow, faucet B does. Phrased differently, if faucet A's valve has pressure, faucet B's valve has no pressure. If faucet A's valve has no pressure, faucet B's valve has pressure.

Let's learn some more from this system. First, notice that faucet B is completely controlled by faucet A. But the reverse is not true: faucet A is in no way controlled by faucet B. Faucet A is controlled by some other faucet, not shown.

Also, notice the control flow of this system is from left to right. This is how we will draw our transistor designs, too. The faucets from off-screen *left* control faucet A. Faucet A, which is *left* of faucet B, controls faucet B.

Now, let's dig a little deeper on this left-to-right flow. Remember I said that in transistor design, we focus on pressure rather than flow: voltage rather than current. Thus far, looking at Figure 4-5 above, we have only focused on two pressures: one at the *valve* of faucet A, and one at the *valve* of faucet B. These are faucet *output* pressures. But, there are two more pressures: one at the *button* of faucet A, and one at the *button* of faucet B. These are faucet *input* pressures. To see this more clearly, look at the picture below:

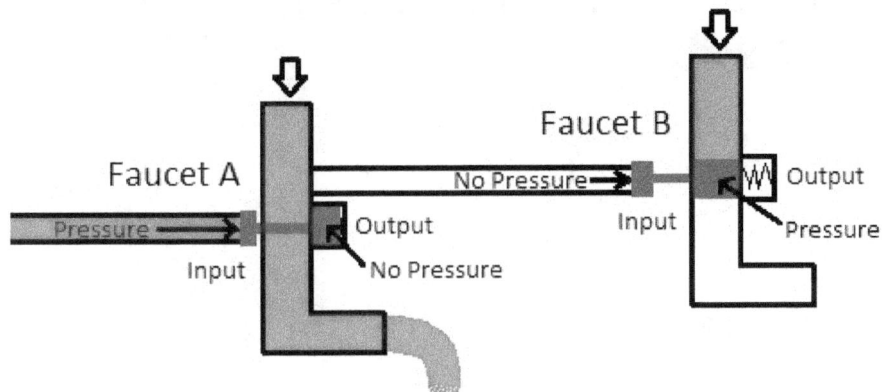

FIGURE 4-6

This is a simplified diagram of Figure 4-5, above. With it, we can focus more clearly on the four pressures. Remember, the *control* flow of the diagram reads from left to right: faucet A controls faucet B. But, the input-to-output *pressure* flow direction is also from left to right. Reading directly from Figure 4-6, faucet A has input pressure, and no output pressure. Faucet B has no input pressure, but has output pressure.

In fact, the output pressure of faucet A controls the input pressure of faucet B. In Figure 4-6, the valve of faucet A is out of the way, the pressure on it is gone, and the water flows along the path of least resistance out of the faucet, so that no water flows across the connector pipe to put pressure on the input button of faucet B.

For the sake of completeness, let's look at how the output pressure of faucet A controls the input pressure of faucet B when the faucets are in their reverse states, as shown in Figure 4-7, below:

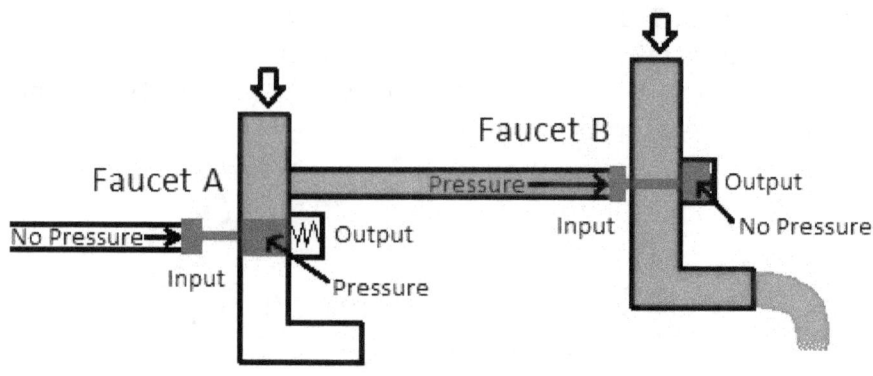

FIGURE 4-7

In Figure 4-7, faucet A's valve is in the way of the water, so it feels pressure. The water can't get through, so it flows across the connecting pipe, applying pressure to the input of faucet B. Faucet A's output pressure causes faucet B's input pressure.

There's still more we can learn from the faucets in Figure 4-7. One faucet acts as the most basic building block of digital logic: the inverter. By definition, an inverter is a logic device that takes the input, reverses it, and places the reversed value on the output. Look at faucet A, Figure 4-7. Its input is "no pressure." The reverse or opposite of "no pressure" is "pressure." Faucet A reverses the input "no pressure" and places "pressure" on the output.

Alternatively, faucet B, Figure 4-7 has "pressure" as its input. Faucet B takes the inverse or opposite of "pressure," placing "no pressure on its own output.

In digital logic, we represent the two binary states as the numbers 1 and 0. We often call 1, "true." We call 0, "false." Since we can associate the presence of pressure with "true," and the absence of pressure with "false," we naturally equate pressure with 1, and no pressure with 0. We do this whether our pressure is from water or from voltage.

Now, instead of describing our faucet as turning an input of pressure into an output of no pressure, we can say that it turns an input 1 into an output 0. Instead of describing our faucet as turning an input of no pressure into an output of pressure, we can say that it turns an input 0 into an output 1.

We can put these results into the form of the table below:

	INPUT	OUTPUT
STATE 1	No Pressure	Pressure
STATE 2	Pressure	No Pressure

(a)

	INPUT	OUTPUT
STATE 1	0	1
STATE 2	1	0

(b)

FIGURE 4-8

The table on the left, (a), of Figure 4-8 summarizes the only two possible behaviors of one faucet. The line listed as STATE 1 corresponds to faucet A's state in Figure 4-7. The line listed as STATE 2 corresponds to faucet B's state in Figure 4-7.

The table on the right, (b), of Figure 4-8 is equivalent to (a), but with binary digits 0 and 1 replacing "No Pressure" and "Pressure," respectively. In this form, the table is called a "truth table," since 1 means true and zero means false. It can also be called a "logic table." From this table, it should be clear that, for an inverter--like our one faucet--a 0 input becomes a 1 output, and a 1 input becomes a 0 output. Now you can see why 0 and 1 are used: they make for a very compact representation, and can stand for whatever we want them to--in this case "No Pressure" and "Pressure."

In digital logic, a logic symbol is often used to represent an inverter. Like the truth table, it is compact, saving us from drawing elaborate transistor circuits or faucets. It also simplifies connecting circuits together, like our faucets controlling other faucets. Look at the picture below:

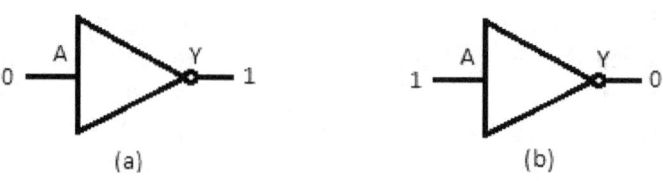

(a)   (b)

FIGURE 4-9

The triangle with a bubble on the right is the logic symbol for an inverter. I drew it twice to show the only two states it can be in. Figure 4-9, part (a) is like state 1 of the truth table in Figure 4-8, part (b): 0 on the input becomes 1 on the output. Figure 4-9, part (b) is like state 2 of the truth table in Figure 4-8, part (b): 1 on the input becomes 0 on the output.

Notice also the direction the triangle is pointing. It is like an arrow, pointing to the right. Remember, I told you that circuit control flows left to right, with inputs on the left and outputs on the right. The shape of the inverter symbol helps us to remember this concept. The input side is often labelled "A," and the output side is often labelled "Y." The bubble at the arrow's tip is a clue that the output is inverted.

If we agree to call the input "A" and the output "Y," we can redraw our truth table for the inverter in an even simpler form than in Figure 4-8. See the picture below, Figure 4-10, part (a), for the simpler inverter truth table.

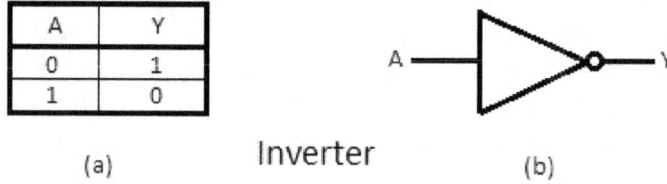

A	Y
0	1
1	0

(a)   Inverter   (b)

FIGURE 4-10

In Figure 4-10, part (b), I also placed a simpler version of the logic symbol from Figure 4-9. Now, with wonderful simplicity, Figure 4-10 displays all you need to know about an inverter. It can be implemented with a transistor, a faucet, or any other binary device you chose, with confidence that they all exhibit these behaviors.

Next, let's redraw Figure 4-7, using logic symbols. Look at the picture below:

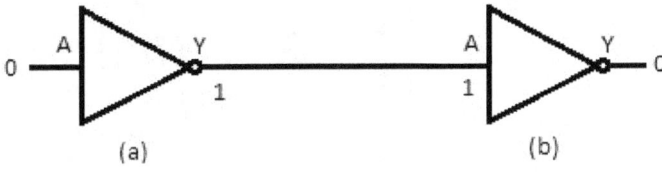

(a)   (b)

FIGURE 4-11

How simple it looks, now. The faucets of Figure 4-7 were two inverters, connected left-to-right. The low pressure (0) coming into inverter (a) was inverted to high pressure. That high pressure was fed to inverter (b). Inverter (b) inverted the high pressure back to a low pressure.

The inverter is just one of the basic building blocks of digital logic. There are only a few building blocks. Yet, when put together, they can create circuits as powerful as a microprocessor. Let's design one more system from water faucets, before we move on to transistors. Look at the picture below:

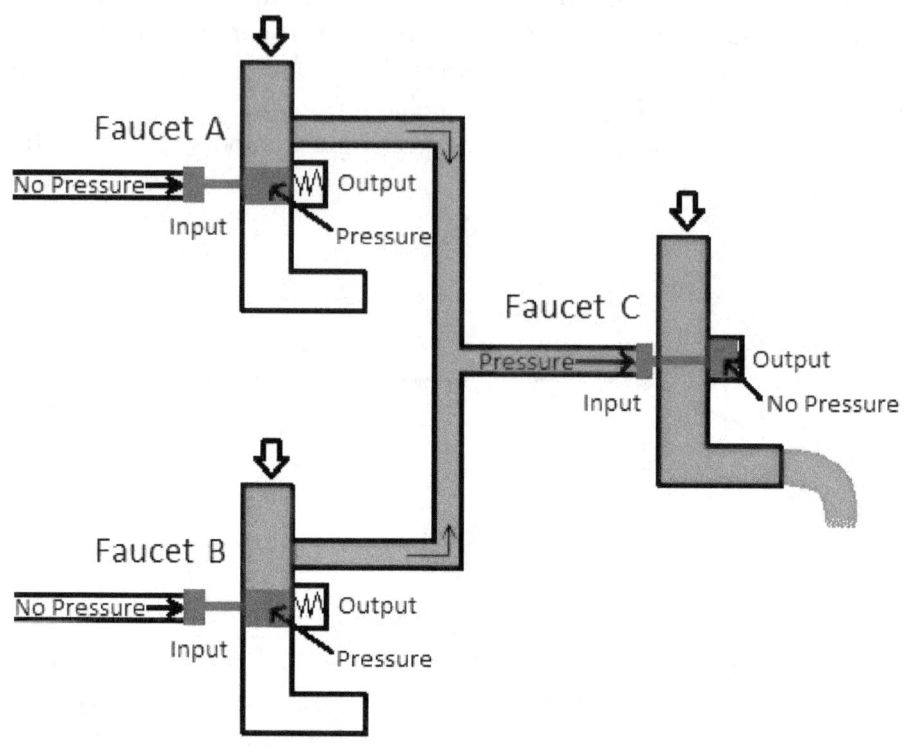

FIGURE 4-12

The first thing you should notice about Figure 4-12 is that faucet C is now being controlled by two faucets, faucet A and faucet B. The inputs to faucets A and B are both at "no pressure." Since they are inverters, the outputs of both faucets A and B are at "pressure." The water and its pressure heads to the input of faucet C, making the input to faucet C, "pressure." Since faucet C is an inverter, its output becomes, "no pressure," and water is free to flow from it.

Now let's change the input to faucet A. Look at the picture below:

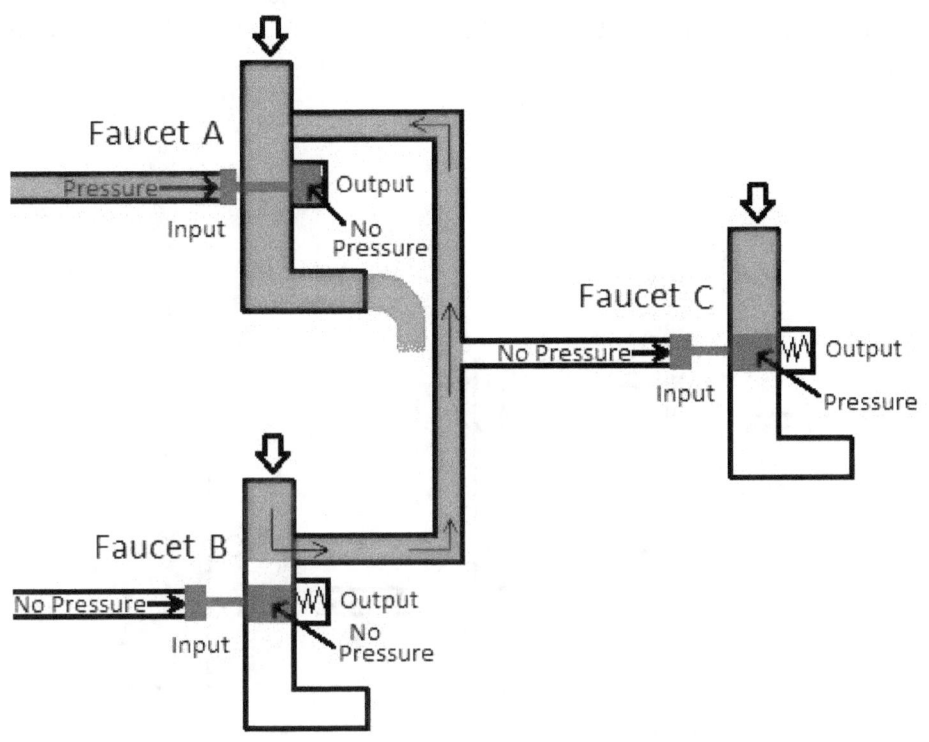

FIGURE 4-13

Figure 4-13 is the same system as Figure 4-12, but now water pressure is applied to the *input of faucet A*. This causes water to stop flowing from *faucet C*, or we would say that the output valve of faucet C now absorbs the water pressure entering the pipe from the top, blocking water flow.

Look closely at the chain of events causes faucet C to turn off. A transistor circuit like this works the same way. Just like before--in Figure 4-6--the water entering faucet A from the top flows out the bottom of faucet A. Thus, the water from the top of faucet A can no longer apply pressure to the input of faucet C. However, the water from the top of faucet B can't apply pressure to the input of faucet C either, because it now has a path of less flow resistance: it too flows out of the bottom of faucet A. In other words, faucet A not only drains itself, it drains faucet B, too. Neither faucet A nor faucet B can turn on faucet C.

Now, look at the picture below:

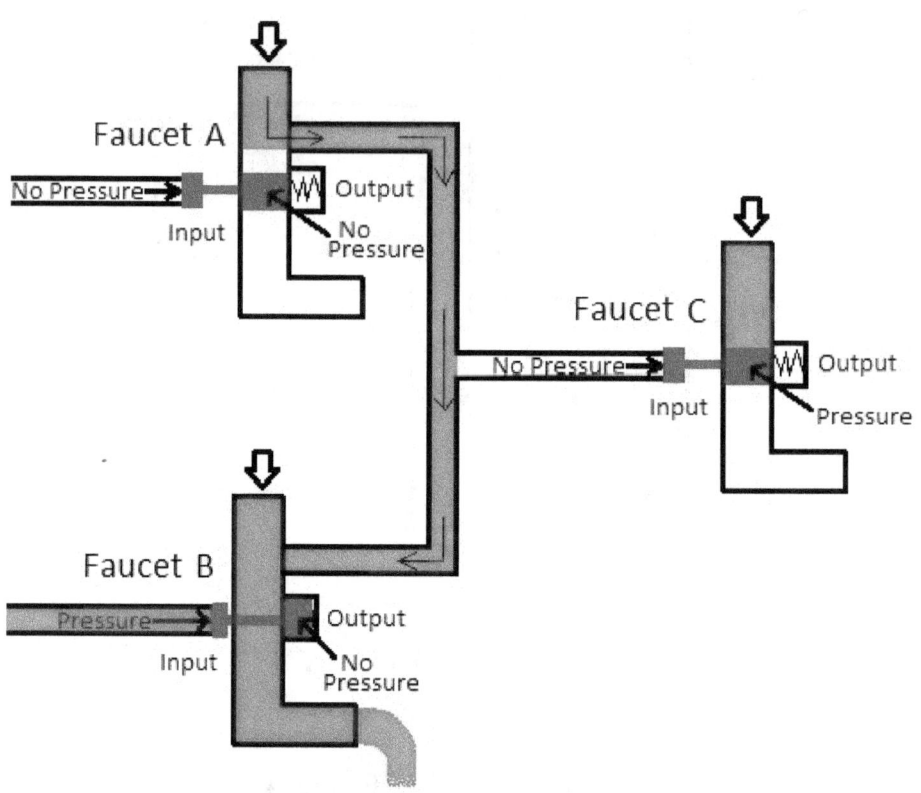

FIGURE 4-14

If you understood Figure 4-13, then you should have no problem with Figure 4-14. The input pressures are now reversed. Instead of input pressure on faucet A and none on B, we have input pressure on faucet B and none on A. But, the results are the same: Faucet C is off--

no water flows--and the pressure on its valve is on. The results in Figure 4-14 are the same as in Figure 4-13, because the system's behavior is the same: one input faucet is on, draining not only itself, but also the other faucet. In this case, faucet B is on, draining not only itself, but also faucet A. Neither faucet A nor faucet B can turn on faucet C. No input pressure is available into Faucet C.

There is only one final possible state for this system. See it in the picture below:

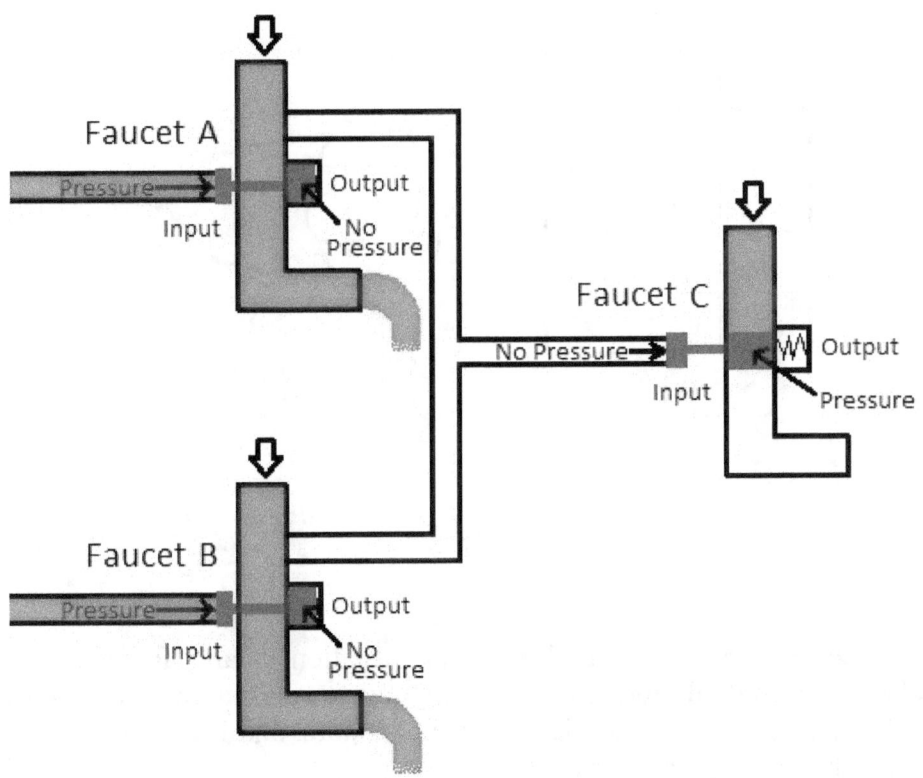

FIGURE 4-15

Figure 4-15 is the fourth possible state for this system. This time, input pressure is applied to both faucets A and B. There's nothing tricky here: no current flows from faucet C; pressure is "on" its output valve. The reason is that, since both faucets A and B are turned on, all their input water runs out of the outlets of their respective pipes. No pressurized water ever reaches faucet C to turn it on.

Figures 4-12, 4-13, 4-14, and 4-15 represent the four possible states of our two-input system. Remember our earlier discussion, where we explained that the way to calculate the number of unique states of a binary system was to take 2 to the power of the number of bits? In this case, there are $2^2=4$ possible input states. This system is our second basic building block of digital logic circuits. It is called an OR gate. See the picture below:

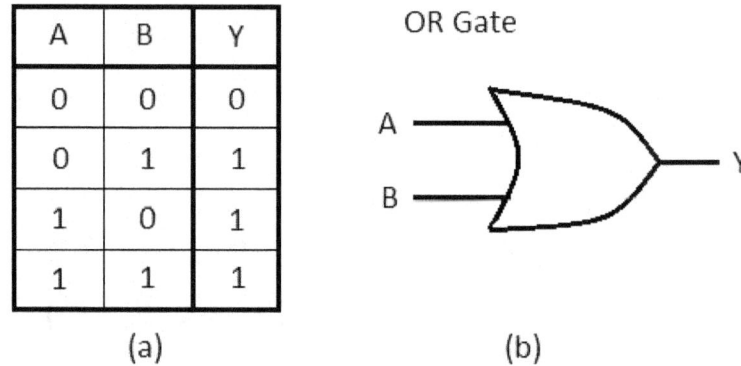

A	B	Y
0	0	0
0	1	1
1	0	1
1	1	1

(a)   (b)

FIGURE 4-16

Figure 4-16 is our simplified representation of an OR gate. It is like our simplified representation of the inverter in Figure 4-10. Figure 4-16 , part (a), is the truth table for an OR gate, and Figure 4-16 ,part (b) is the logic symbol for the OR gate.

Let's first try to understand what part (a) is telling us. Remember that 1 represents "true", as in "the pressure is truly on." 0 represents false, as in "the pressure is not on." The A and B columns represent the states of the input pressures, the inputs to faucet A and faucet B. The letter Y--not C--is traditionally used as the name for the state of the output pressure, in this case the output of faucet C. The four rows represent the four Figures: 4-12, 4-13, 4-14 and 4-15. How simple and compact! A whole diagram gets replaced by a simple row of a truth table. Those confusing intermediary, internal pressures that coupled the outputs of faucets A and B to the input of valve C do not appear. We really don't need to know about them when designing at the logic gate level.

Part (a) helps clarify what we should have noticed when studying Figures 4-12 through 4-15: that, as long as at least one input pressure is on, the output pressure is on. It is called an OR gate because, if the A pressure input *or* the B input, *or* both inputs are on, then the output pressure is on.

Figure 4-16, part (b) is the logic symbol for an OR gate. Notice how the left side of the symbol is curved, and how the top and bottom sides are curved and come to a point at the right. This helps us remember the left-to-right flow in circuit design and analysis. To design, we don't need to know what's inside (b): it could be big pipes, it could be little transistors. We design with logic symbols.

Let's do a design, now. So far, you know two logic gates: the inverter and the OR gate. Let's design a system to monitor your toilet. If someone flushes the toilet and immediately leaves the house, you could get water damage to your house if the bowl *or* the tank overflows. Hint: I just used the word, "or." See the picture below:

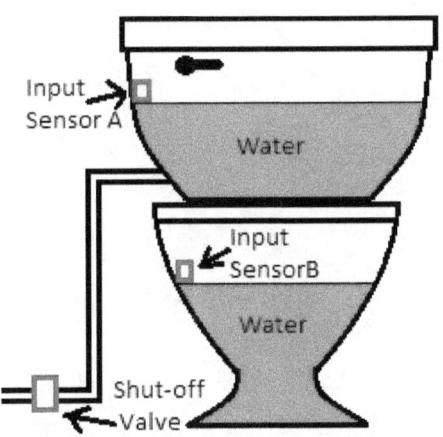

FIGURE 4-17

Let's forget about voltages or water faucets inside logic gates, for now. Let's concentrate on what 1's and 0's will stand for. Let's say that the inputs to our design are two sensors: one that monitors the height of the water in the toilet bowl, and one that monitors the height of the water in the toilet tank. Normally, they are in the logic 0 state, meaning the water is *not* too high (false). If something goes wrong, a sensor goes into the logic 1 state, meaning the water *is* too high (true).

The output of our system is a water shut-off valve. Normally, it is in the logic 1 state, meaning water is allowed (true) to come into the toilet from the water supply. If something goes wrong, it must go into the logic 0 state (false), meaning water is not allowed to come into the toilet from the water supply.

If you like, take a break and try designing this system yourself. Put the names of two input sensors on the left, the name of the one output shut-off valve on the right. In the middle, put whatever combination of logic gates you need to create the control system. Don't get hung up on the sensors or shut-off valve. Just treat them like sources or destinations of 1's and zeroes. The answer is in the picture below:

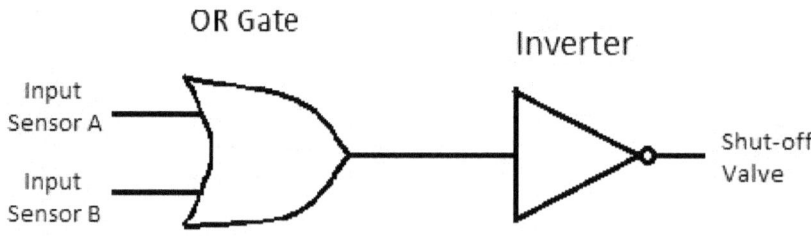

FIGURE 4-18

The OR gate--as I hinted earlier--is almost enough for this design. Remember from the truth table in Figure 4-16 for the OR gate that its output will be low if both inputs are low; its output will be high if either input or both inputs are high. This is the opposite of what we require. We want the output to be high--turning on water flow--when both inputs are low--indicating no faults. We want the output to be low--turning off water flow--when either or both inputs are high--indicating fault(s). We want to invert (make opposite) the output of the OR gate. We do this by attaching an inverter to the output of the OR gate. Look at the picture below:

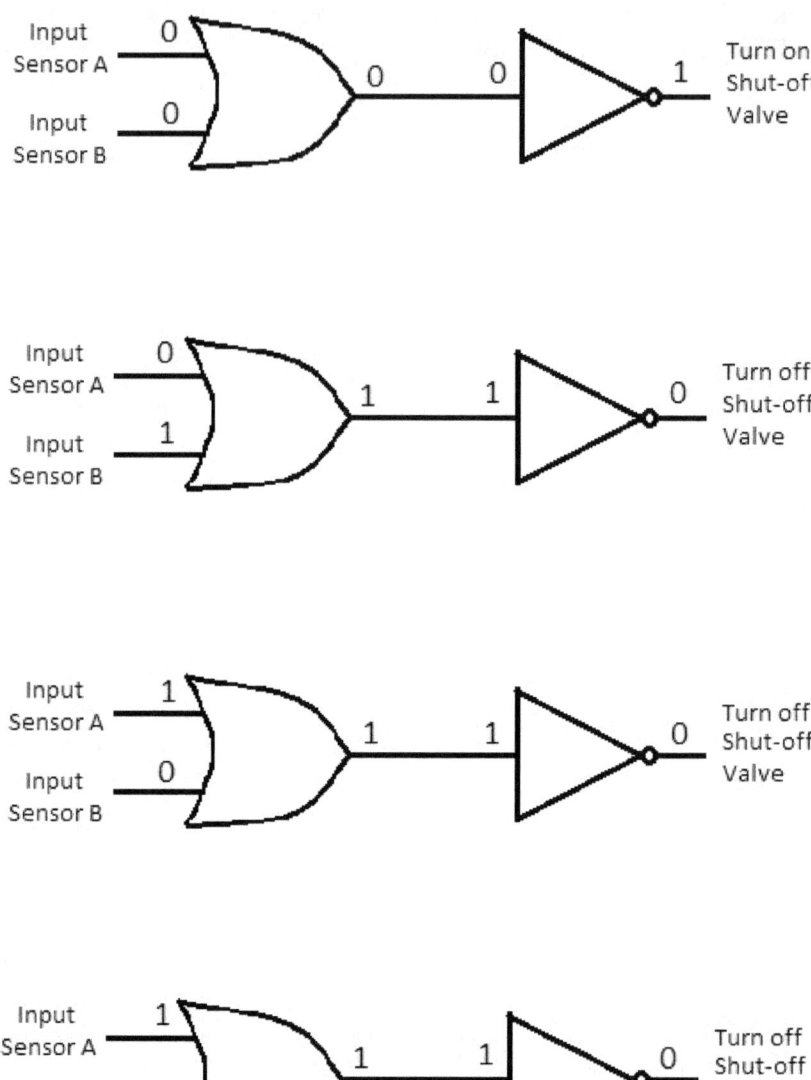

FIGURE 4-19

The four rows of Figure 4-19 show the 4 possible states for our system. Any input of 1, on the far left side, means water trouble. The output, on the far right, of that state must go to zero, to prevent water damage. From this figure, you can see that the inverter flips the OR gate's output, to make it work correctly.

Let's compare the OR gate truth table we encountered in Figure 4-16, part (a), with the truth table we can build from Figure 4-19. See the picture below:

OR Gate   NOR Gate

A	B	Y
0	0	0
0	1	1
1	0	1
1	1	1

(a)

A	B	Y
0	0	1
0	1	0
1	0	0
1	1	0

(b)

FIGURE 4-20

Part (a) of Figure 4-20 is the truth table for the OR gate. Part B of Figure 4-20 is built from the four rows of Figure 4-19. The A and B columns of part (b) are from the far left side of Figure 4-19: the inputs. The Y column of part (b) is from the far right side of Figure 4-19: the output. This makes it easy to see how the Y output column is inverted in Figure 4-20 part (b) compared to part (a).

It turns out that the circuit we just designed is our third fundamental building block of digital electronics: the NOR gate. NOR stands for "Not OR," in reference to the fact that each of the possible output states of the OR gate are not what they used to be. To make an OR gate, we simply need to add one extra inverter--an extra water faucet or transistor--to the right end of the three-inverter OR gate we already created, as in Figure 4-15.

The picture below summarizes the NOR gate's truth table and logic symbol:

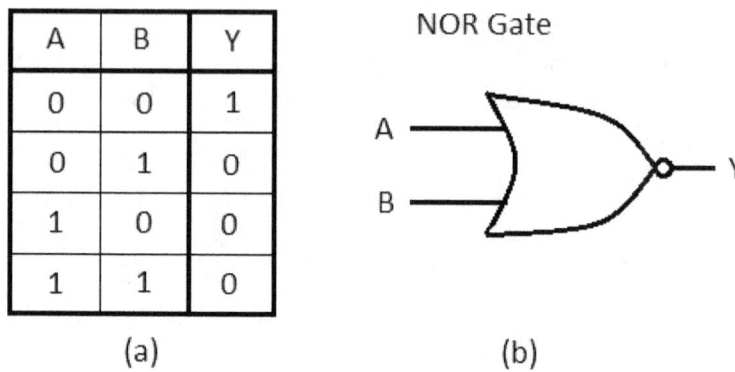

A	B	Y
0	0	1
0	1	0
1	0	0
1	1	0

(a)   (b)

NOR Gate

FIGURE 4-21

Notice, in Figure 4-21, part (b), that the logic symbol for the NOR gate looks just like the logic symbol for the OR gate, except that the NOR gate has a bubble on the output. The bubble means that the output is inverted from the OR gate. This is a symbolic way of showing the same thing we discovered in Figure 4-20, comparing the Y output columns of the OR and NOR gates. Remember that the NOR gate logic symbol represents a physical system with four faucets inside: three from an OR, and one from an inverter.

CHAPTER FIVE
SWITCHES

We have spent enough time on faucets. Our goal is to teach you how these OR and NOR gates, inverters, and microprocessors are implemented with transistors. But, before we do that, we must pick up the story of electronics where we left off.

In our exploration of electronics we found that the "pressure" of voltage produced by a battery is caused by positive charge forced together on one terminal, and negative charge forced together on the other. This resultant pressure is comparable to the water pressure produced by a water pump, caused by a spinning propeller.

We next found we could make current flow--make electrons move--from the negative battery terminal to the positive terminal of the battery. We did this by connecting a semi-conductor, like a resistor, across the battery terminals. Let's look at resistance for a moment. See the picture below.

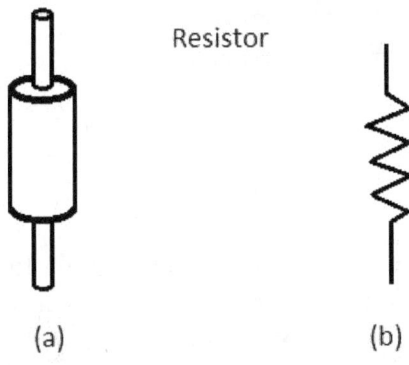

Resistor

(a)   (b)

FIGURE 5-1

One kind of resistance, or semiconductor, is called a resistor. It is an electronics component specifically designed to control the amount of current flow. The one in Figure 5-1, part (a), is about a centimeter long, and costs a penny. The electrical symbol of resistance is the jagged line shown in Figure 5-1, part (b).

When we buy a resistor, we choose how much we want it to limit current flow. The unit of measurement of resistance is ohms, symbolized by the omega symbol: Ω. If we buy a resistor with more ohms (more resistance), less current flows. We can buy resistors in units of less than an ohm up to units of greater than a million ohms.

We can actually calculate how much current will flow with the equation called Ohm's Law, which states that I=V/R. "I" is the letter symbol for current, in units of amperes, "V" is the letter symbol for voltage, in units of volts, and "R" is the letter symbol for resistance, in units of ohms. For example, if we connect a 6 ohm resistor across a 12 volt battery, then 2 amperes of current will flow, since I=V/R=12/6 =2 amps. See the picture below.

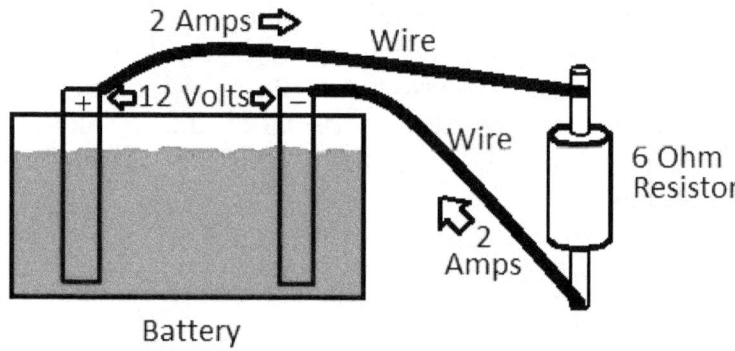

FIGURE 5-2

Figure 5-2 is the way we should have built figure 3-2, since it contains a resistor to limit current flow. There are a few things you should notice. First, we have connected the resistor to the battery with two wires. Second, the voltage appears across the battery terminals, and is applied across the resistor. Third, the current flows through the two wires, through the resistor, and through the battery. Forth, even though the electrons flow from the negative battery terminal to the positive terminal, we usually draw the "conventional" current flow the other way: from the positive terminal to the negative. This convention is due to an early bad guess by investigators who assigned this direction before we knew about electrons.

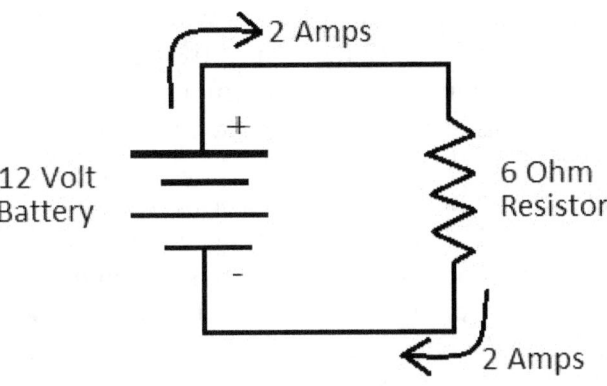

FIGURE 5-3

Figure 5-3 is the schematic diagram of the circuit in Figure 5-2. A schematic diagram is an electronics shorthand method of representing a physical circuit. It uses schematic symbols in place of pictures of actual components. I have introduced these symbols in previous figures. Given Figure 5-3, an electronics technician or engineer can build the circuit drawn in Figure 5-2.

Let's look at another schematic diagram. See Figure 5-4, below:

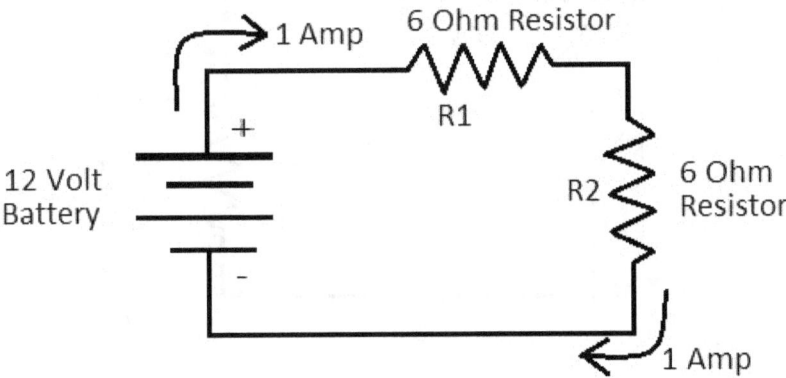

FIGURE 5-4

Compare Figure 5-4 to Figure 5-3. The first thing to notice is a second resistor. If you were to build this circuit, you would connect a wire from the positive terminal of the battery to one end of the resistor labeled R1. You would connect a second wire from the other end of resistor R1 to one end of resistor R2. You would connect a third wire from the other end of resistor R2 to the negative terminal of the battery. You could solder the wire connections, or use wires terminated by alligator clips or spring-loaded clips. The wires could even be ribbon-like copper traces on a circuit board.

The second thing to notice is that half as much current flows now, compared to Figure 5-3. This is because the same 12 volt battery now has to work twice as hard, forcing electrons through twice as much resistance as before: 6 ohms + 6 ohms = 12 ohms. Twice as much resistance to the same voltage leads to half as much current. Ohm's Law predicts this:

$I = V / (R1 + R2) = 12 / (6 + 6) = 12 / 12 = 1$ amp.

Let's redraw Figure 5-4 as Figure 5-5, with resistor values changed:

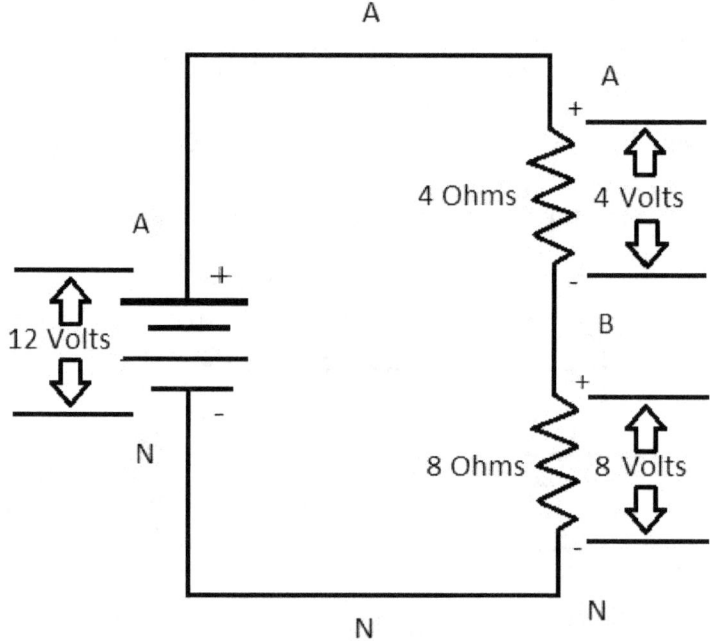

FIGURE 5-5

In Figure 5-5, we are revealing something strange. There is certainly nothing odd about labeling the voltage across the 12 volt battery as being 12 volts. But strangely, we are claiming that 4 volts appears across the 4 ohm resistor, and 8 volts appears across the 8 ohm resistor. Are we trying to claim that the resistors, like the battery, can create voltage?

Absolutely not. Figure 5-5 reveals just the opposite: in a sense, the resistors take away voltage. Figure 5-5 exhibits for us the Law of Conservation of Energy. In this case, all the energy put into the system is exactly equal to all the energy that comes out of the system as work.

It would be hard for you to come to this conclusion, because traditional electronics terminology hides the truth. You look at the circuit of Figure 5-5 and see three voltages. But these voltages are not the same. It would be nice if they were labeled differently, but they are not: they are all called, "voltages," and measured in units called, "volts."

In truth, the voltage across the battery is one kind of voltage, and the voltage across the resistors is another kind of voltage. The voltage across the battery is input *energy* (per coulomb of charge). The voltage across the resistors is output *work* (per coulomb of charge). Now when you look at Figure 5-5, you can see the Law of Conservation of Energy in play. The input energy is 12 volts, across the battery. This exactly equals the output energy (work) of 4 volts plus 8 volts equals 12 volts across the resistors.

The little plus and minus signs in Figure 5-5 are a tradition meant to flush out this hidden difference. If you trace your finger in a clockwise direction around this circuit, you will find that the order of the plus and minus signs are opposite when encountering a battery, compared to encountering a resistor. As your finger goes clockwise, when your finger enters a battery, it will touch a negative sign, and when it exits a battery, it will touch a positive sign. Still going clockwise, when your finger enters a resistor, it will touch a positive sign, and when it exits a resistor, it will touch a negative sign.

Still focusing on Figure 5-5, notice the labels A, B and N. In an electrical circuit, we usually pick a neutral point, which I have labelled, "N." The neutral is a point of reference for all our voltage measurements. We choose it to be our "zero volts" point. Once we pick our neutral, all our voltage measurements are relative to neutral. There is no absolute zero voltage point, only a relative zero voltage point. Remember, the battery has pressure on both ends, negative pressure on the anode, and positive pressure on the cathode. We traditionally pick the negative battery terminal as neutral. The wire leading to the bottom resistor is also neutral.

Moving up the left side of Figure 5-5, from N to point A, the battery causes our circuit to go from zero volts up to +12 volts. We now have +12 volts relative to neutral. Sometimes we say we have gained 12 volts of electrical potential.

You may remember the concepts of potential and kinetic energy from high school physics class. At point A, our circuit now has the potential energy to do 12 volts worth of work. But what is the work? The work is forcing electrons through 4 + 8 = 12 ohms of resistance.

Look now at point B in Figure 5-5. At this point, the 12 volts of electrical voltage potential has accomplished some work: it has forced electrons through the top, 4 ohm resistor. We measure that work as 4 volts across the 4 ohm resistor. We say we had a voltage drop of 4 volts, because we began with 12 volts, but did 4 volts of work, so we have 12 - 4 = 8 volts left to force current through the 8 ohm resistor. When we measure the 8 volts across the 8 ohm resistor, we are measuring the work done by the battery to get the current through the 8 ohm resistor. We started with 12 volts of energy, completed 12 volts of work, and have arrived back at neutral (N), back at zero energy.

But, why is there 4 volts of work done across the 4 ohm resistor, and 8 volts of work done across the 8 ohm resistor? Well, we know that the total work has to equal 12 volts, because the total energy from the battery is 12 volts, and the Law of Conservation of Energy says the work out must equal the energy in. Very simply, it's twice as much work for the battery to force electrons through an 8 ohm resistor than it is to push electrons through a 4 ohm resistor. So, we measure twice as much voltage, 8 volts, across the 8 ohm resistor, than the 4 volts we measure across the 4 ohm resistor.

I chose 4 ohms and 8 ohms for my convenience, to arrive at 4 and 8 volts. But, if the upper resistor had been 12 ohms and the lower resistor had been 24 ohms, we still would have measured 4 volts across the upper resistor and 8 volts across the lower resistor. The total has to be 12 volts, and it's twice as hard to get current through a 24 ohm resistor as through a 12 ohm resistor, so 4 volts must be measured across the 12 ohm resistor and 8 volts must be measured across the 24 ohm resistor. (Of course, less current will flow in the circuit with the 12 and 24 ohm resistors than in the circuit with the 4 and 8 ohm resistors.) As an analogy, it's twice as much work for you to lift a 24 pound weight as it is for you to lift a 12 pound weight.

Figure 5-5 might remind you of physics concepts of pushing a rock up a hill. The left side of Figure 5-5 is uphill. The right side is downhill. You have energy from the food you eat. (The battery has energy in its chemical structure.) You do work by pushing a rock up a hill. (The battery does work by separating charge.) The rock now has potential energy. (The positive and negative ions have electrical potential energy.) When the rock falls down the hill, it releases energy as kinetic, work energy. (When the electrons go down through the resistors, they release energy as work-done energy).

When the rock falls back down the hill to its starting point, it no longer has the energy you put into it. (When the electrons get back to neutral, they no longer have the energy the battery put into them. The battery must put energy back into them to keep current flowing.)

You may think we have gotten off-track. Yes, this is electronics. But what does all this have to do with digital, or binary systems? Right now, we are going to make one change in this system to turn it into binary. Look at the picture below:

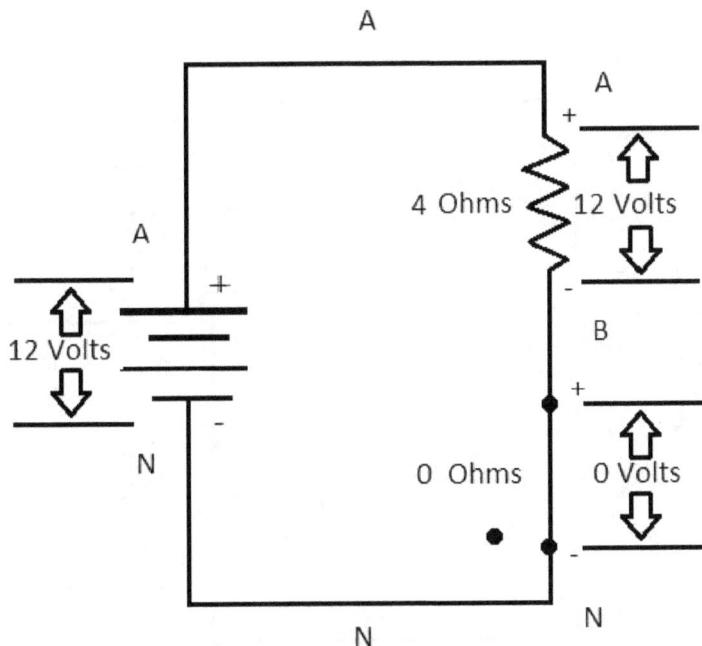

FIGURE 5-6

Figure 5-6 is just like Figure 5-5, but the bottom, 8 ohm resistor has been changed to a 0 ohm resistor. Rather than draw the jagged resistor symbol, I have drawn a wire. A wire has 0 ohms of resistance--or no resistance to current flow--because it is a conductor. In fact, I have drawn the symbol for a closed switch. You have used switches thousands of times in your life. A closed switch turns on a light or brews the coffee. A closed switch is just a wire.

We need to make sense of the voltage measurement across the wire, the closed switch. The battery is, like before, 12 volts. So, the voltage drops across the resistor and the wire must add up to 12 volts. They do: the 12 volts across the resistor plus the zero volts across the wire equal 12 volts. Why is there 0 volts across the wire? Because, it takes no work for the battery to get current through a wire. A wire does not resist current flow: it is not a semiconductor, it is a conductor. All the work the battery does getting current through the circuit is expended on the one resistor.

Now, let's open the switch. Look at the picture below:

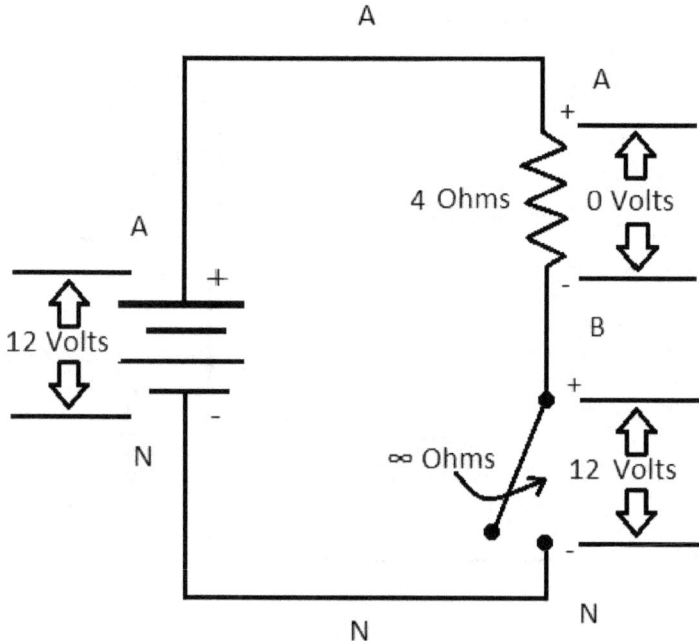

FIGURE 5-7

The only thing we have done in Figure 5-7--compared to Figure 5-6--is opened the switch. Remember, the switch is the component on the bottom-right of our figure. When we open a switch, a light goes off, or coffee stops brewing. We effectively remove a wire from a circuit. Actually, one end of the wire is pushed to the side, just like pictured in Figure 5-7. The circuit is no longer a circuit: current can't "circulate."

Such a simple change causes some strange results in Figure 5-7. First, the resistance across the switch is now infinity! There is now, effectively, no component where the wire of the closed switch used to be. The only thing between the two contacts--the dots in the figure that are points of contact to the external wires of the circuit--is air. And, air is an insulator. Insulators have infinite resistance.

Next, the voltage measurements in Figure 5-7 have reversed, compared to Figure 5-6. Zero volts are now across the resistor, and 12 volts are now across the open switch. This result is often not what people expect. People expect to find zero volts at the switch if "nothing" is there.

Instead, we find 0 volts across the resistor, because no work is being done there. Since this is an "open" circuit, no current is flowing. If no current is flowing through the resistor, the battery is doing no work there.

Next, let's look at the 12 volts across the open switch. As we went clockwise from point N in the circuit, through the battery, to point A, potential energy rose to +12 volts. Since no volts dropped across the resistor (no current flow means no work done) the voltage at point B must still be the same as at point A: 12 volts. Since all voltages are relative to neutral, that means we have 12 volts across the open switch.

Another way to look at this is by comparing it to the "water faucet off" figure. In the "off" water faucet, no current flows, so all the water pressure is applied to the valve. Similarly, in Figure 5-7, since no current flows, undiminished voltage is applied to the top of the switch (point B), relative to the bottom (point N).

You may choose to view the right side of Figure 5-7 to be like two voltage drops, $0 + 12 = 12$ volts, adding up to the voltage rise on the left side of Figure 5-7. We did this in Figures 5-5 and 5-6. Or, you may choose to see no voltage drops: no work has been done, so everywhere we look--other than neutral--we find potential energy, or 12 volts.

Comparing Figures 5-6 and 5-7, we see good reason for leaving electrical work to trained people. Figure 5-7 could be the light switch in your house; the resistor could be a light bulb; and the voltage could be a dangerous 120 volts AC. An untrained person might think that, since the switch is off and the light is off, then the electricity is off, and it's safe to touch the switch contacts. WRONG! That person would be part of a live circuit, and could get a fatal shock. Counter-intuitively, it is safer to touch across the closed switch in diagram 5-6, since no voltage appears across it. Don't do this either: something else could be wrong with the circuit, so it could still fatally shock you.

At this point, we have a binary electrical system. The two states are, physically, 12 volts and 0 volts. Like with the faucet system, "on" and "off" are backwards. When the switch is on--making current flow and a light turn on--the voltage across the switch is "off," or 0 volts. When the switch is off--making current stop and a light turn off--the voltage across the switch is "on," or 12 volts.

CHAPTER SIX
TRANSISTORS

Unfortunately, the switch-based binary system is not good enough to satisfy our needs. It requires a human to flip the switch. That's why we need the transistor. We need an electronic component that acts like a switch, but has an input side that controls whether the switch is on or off. And, that input side must be able to be controlled by the outputs of other transistors. Let's look now at a transistor circuit. See the picture below:

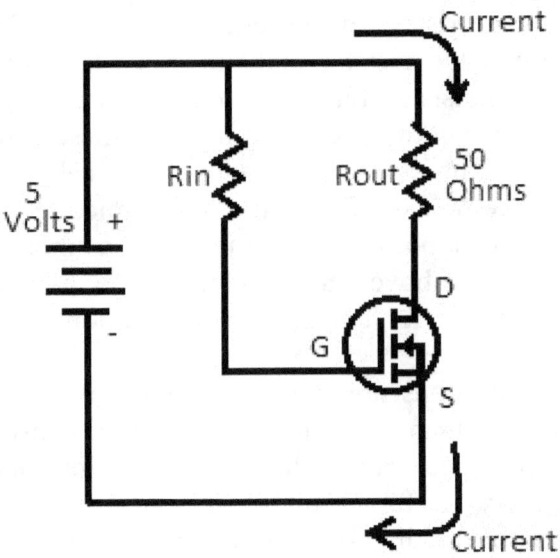

FIGURE 6-1

Before we proceed, remember out goal. We want an electronic device that acts like an electrical switch (Figures 5-6 and 5-7), but one that can be controlled by other devices of the same kind (like the faucets of Figures 4-4 through 4-7).

In Figure 6-1, we introduce a transistor. Its electronic symbol is in the lower right-hand corner: a circle containing an alien-looking symbol. It is a three-terminal device, with the connections named S, G, and D, for source, gate, and drain.

First, let's get some physical exercise. Figure 6-1 separates neatly into two, interrelated paths. Use your finger to trace one path. Start at the + terminal of the battery. Move clockwise: up, right, then down through the resistor labeled Rout. Keep going through the transistor, entering the drain (D), and exiting the source (S). Go down, then left, then up to the negative terminal of the battery. This is the output path of our circuit.

Next, let's trace the input path. Again, start at the + terminal of the battery. Move clockwise: up, right, then down through the resistor labeled Rin. Keep going through the transistor, entering the gate (G), and exiting the source (S). Go down, then left, then up to the negative terminal of the battery.

As you trace each of these two paths, try to see that each one resembles the kinds of circuits we saw in Figures 5-5, 5-6, and 5-7. Imagine the voltage rising on the left, then falling on the right, as it progresses through the resistor and the transistor. Notice how both paths involve the transistor, but through different entry points. We will have much to say about the interrelationship of these two paths.

The kind of transistor I have chosen is called an N-channel enhancement-mode MOSFET (metal-oxide semiconductor field effect transistor). I chose it because it is the one that most resembles the water faucet model above, and because it is commonly used in microprocessors due to its low power consumption and fast switching speed.

Here's how this transistor works. Applying a voltage at the gate that is more positive than the voltage at the source causes *conventional* current to flow from the drain to the source (electrons actually flow the other way, but try to forget that). The greater the gate-to-source voltage, the greater the drain-to-source current. That's all you need to know for a basic understanding of this transistor.

So, what can we do with transistors? Well, for one thing, we can make a voltage amplifier. A small audio sine wave across the gate-to-source causes a sine wave of current from drain-to-source, which causes a large voltage sine wave across a suitably large Rout resistor.

But we care here about the use of the transistor not as an amplifier, but as a switch. Here's what makes transistors different, and a little strange. When we go back to our basic electronics theory of Figure 5-3 (one battery across one resistor), we find that current is being controlled by our choice of voltage and resistance. Once we chose a 12 volt battery and a 6 ohm resistor, the amount of current was determined. Ohm's Law said we would get: $I=V/R=12/6=2$ amps of current (Review: I stands for current, V stands for voltage, R stands for resistance).

But, the Rout resistor does not control the amount of current flow through a transistor. Look at Figure 6-1. +5 volts from the battery is across the combined 50 ohm resistor and the drain-to-source of the transistor. And yet, the 50 ohm resistor does not affect the current through that branch. If it was a 100 ohm resistor, the same amount of current would flow! This upends what a beginner in electronics is taught: that more resistance causes less current.

We have to come up with a new name for the drain-to-source path of the transistor. We call it a dependent current source. It's kind of the opposite of a battery. A battery produces a (somewhat) constant voltage, regardless of what current comes out of it. A current source produces a (somewhat) constant current, regardless of what voltage appears across it.

So, the current from drain-to-source is determined. What determines it? I already told you: the voltage across the gate-to-source. The drain-to-source current is the same current that goes through Rout. Yet, the value of Rout does not control the amount of current that goes through itself! The drain-to-source current controls Rout's current, and the drain-to-source current is itself controlled by the gate-to-source voltage. If you want to be really fancy, you can call this MOSFET dependent current source a voltage-controlled current source, since the gate-to-source voltage controls the drain-to-source current.

Since the gate-to-source voltage is the controller, what is the value of the gate-to-source voltage? Well, the +5 volts from the battery goes across the combined Rin resistor and the gate-to-source of the transistor. But you need to know this: the resistance of the MOSFET transistor's gate-to-source path is nearly infinite, like an open circuit. If you look back at Figure 5-7 and remember how voltages behave in an open circuit, you will see that all the *+5 volts appear across the gate-to-source*, and zero volts appear across the Rin resistor.

This is one reason that this transistor is a good choice for microprocessors that have millions of transistors packed together. No current flows through Rin or through the gate-to-source of the MOSFET. No power, hence no heat is generated down that path. Control of the output is maintained, not by input current, but by the electric field generated by the input voltage. (We explained this electric force field earlier in this book.)This electric field is inside the MOSFET. The gate-to-source voltage controls the electric field intensity, and the electric field intensity controls how much current is allowed to flow from drain-to-source. That's why this device is an FET: field effect transistor.

To reiterate: +5 volts appears across the gate-to-source of the transistor. From this, how do we know how much current flows from drain-to-source? We look it up in the data sheet for the particular MOSFET we bought.

Let's say we find that our model MOSFET produces .1 amps from drain-to-source when +5 volts are present from gate-to-source. This means that .1 amps also flows through the 50 ohm, Rout resistor. From this, we can calculate the voltage drop across the 50 ohm resistor. We use algebra to rearrange Ohm's Law, I=V/R, so that it becomes V=IxR.

This form of Ohm's Law beautifully describes how a voltage drop, V, is the work done (per coulomb of charge) by the energy source in getting current through a resistance. In V=IxR, if I is a fixed value, then increasing R means more work was done in getting that same I (current) through a bigger resistance. On the other hand, if R is a fixed value, then increasing I means more work was done pushing more current through the same resistance.

Using this equation, we find that the voltage drop across the 50 ohm resistor is: V=IxR=.1x50=5 volts. I chose these values on purpose, because I wanted 5 volts. You'll see why in a moment. (Aside: in a real world circuit, I would choose a bigger resistor value, to reduce power consumption, but that is an advanced topic that would ruin the clarity of this tutorial.)

Look now at the picture below:

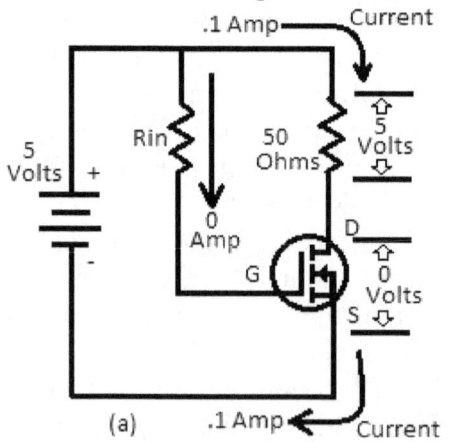
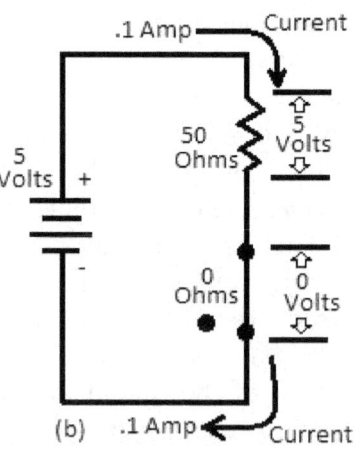

FIGURE 6-2

On the left I have redrawn Figure 6-1 with all the voltages and currents. On the right I have redrawn Figure 5-6, slightly modified to be more equivalent to the circuit to its left. We just concluded that we have a 5 volt voltage drop across the 50 ohm resistor. If we started with a 5 volt voltage rise from the battery, and used 5 volts of work to get current through the 50 ohm resistor, we have 0 volts of work let to get the current through the drain-to-source of the transistor. This means that the drain-to-source of the transistor, with zero volts across it, is acting like a wire or a closed switch: a component with no resistance, causing no voltage drop.

Compare Figure 6-2, part (a), with Figure 6-2, part (b), to see this more clearly. All the voltages and all the currents in the output path align exactly. The transistor is behaving like a closed switch. It may have taken us many paragraphs to explain the process, but the result is simple. Notice also that the input path through Rin to the gate seems to disappear, since no current goes down it.

The path through Rin is important, though. Let's change that path. Look at the picture bellow:

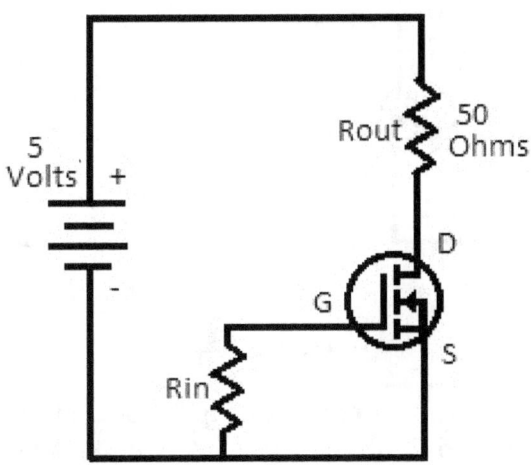

FIGURE 6-3

Compared to Figure 6-1, the only change is that one end of Rin is now connected to neutral: to the negative end of the battery.

Remember, neutral is the 0 volts point, the reference point from which we make all our voltage measurements. We now have no voltage difference between the gate and the source of the transistor. If you trace your finger from the negative of the battery--0 volts--and go up through Rin, out the top of Rin, turn right, enter the gate of the transistor, then come out the source of the transistor, you wind up where you started from: 0 volts.

Reiterating: there is no voltage difference between the gate and the source of the transistor. The drain-to-source current of the transistor is controlled by the amount of gate-to-source voltage. Less gate-to-source voltage produces less drain-to-source current. And--the specification sheets for this transistor tell us--no voltage difference between the gate and source of the transistor means no current flows from its drain to its source: 0 amps. If there is no drain-to-source current, it is effectively an open circuit. Look at the picture below:

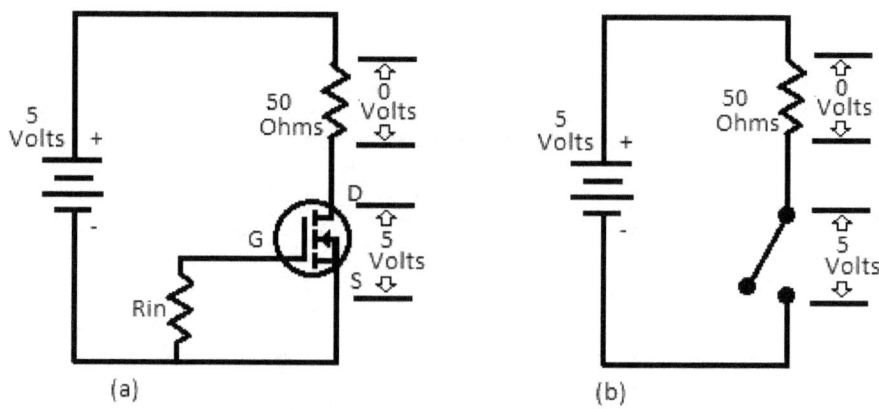

FIGURE 6-4

Figure 6-4, part (a), is a repeat of Figure 6-3. Figure 6-4, part (b), is a repeat of Figure 5-7, an open switch circuit. If you compare the voltages in these two circuits, you will notice that their voltages exactly correspond. You will also notice that their currents exactly correspond: no current flows anywhere. You will conclude that the MOSFET transistor circuit with 0 volts input at the gate behaves exactly like an open switch.

Let's summarize, then. A MOSFET transistor circuit with 5 volts input at the gate behaves exactly like a closed switch (Figure 6-2). A MOSFET transistor circuit with 0 volts input at the gate behaves exactly like an open switch (Figure 6-4). Furthermore, the MOSFET switch is controlled by input voltage, rather than a human hand flipping the switch. The input voltage can be from other MOSFET transistors.

We were getting surprisingly far with our building our binary logic system from the ground up with water faucets. Let's place a faucet side-by-side with our MOSFET transistor circuit. See the picture below:

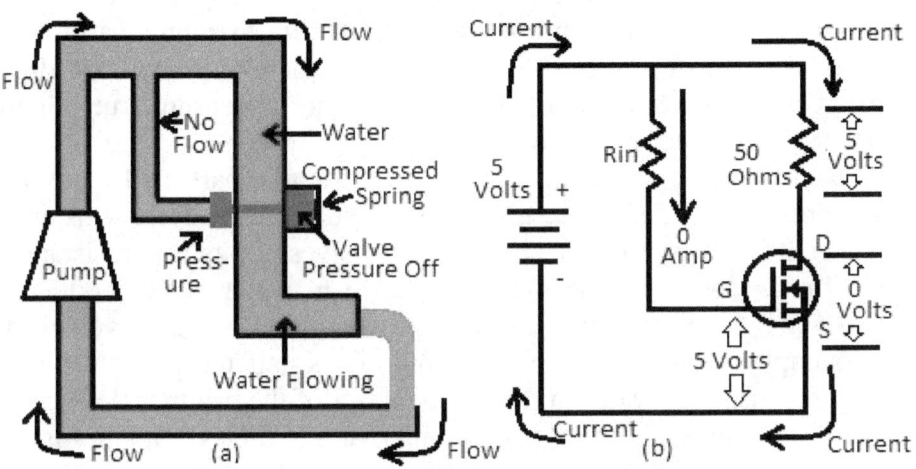

FIGURE 6-5

Close examination reveals how similar the two circuits are. My hope is that your understanding of water systems you have been using since you were a child will help you understand the less intuitive transistor circuit.

Let's start with the flow. I modified the water system a little, to have the water circulate instead of coming from a reservoir. (Otherwise, you would deplete this system when you drink the water.) If you compare the water flow path with the current flow path, you will find that they circulate identically. Notice how, in part (a), the water that is applying pressure to the valve's button is *not* flowing (moving). Nevertheless, it is a conduit for the pressure from the pump. The same phenomena happens in (b). Even though 0 amps of continuous current flows down toward the gate of the transistor, the redistribution of electrons on the source and gate caused by the battery creates the gate-to-source voltage.

Let's now compare the water pressure to voltage. The pump is the source of water pressure, as the battery is the source of voltage. In Figure 6-5, part (a), pump pressure (transmitted through the water) applies pressure to the valve's button, turning on water flow through the faucet. In Figure 6-5, part (b), battery voltage (causing a redistribution of electrons on the source and gate) applies the voltage that turns on current flow through the output drain-to-source path.

Also, notice in Figure 6-5, part (b), that there is no voltage drop across the drain-to-source of the transistor. The same holds true in part (a): there is no water pressure on the valve.

The difference you might find between the two parts of Figure 6-5 is the presence of resistance in part (b). But, there really is resistance to water flow in part (a). The pipes themselves act as this resistance. The skinnier the pipe, the more resistance to water flow, producing less water flow: Flow_rate=1/4xπx(pipe_diameter)^2 x(water_velocity). For example, replacing a pipe with one that's half the original pipe's diameter causes the water to flow at a quarter of the original flow rate.

Let's now look at the no-flow counterparts to the systems in Figure 6-5. See the picture below:

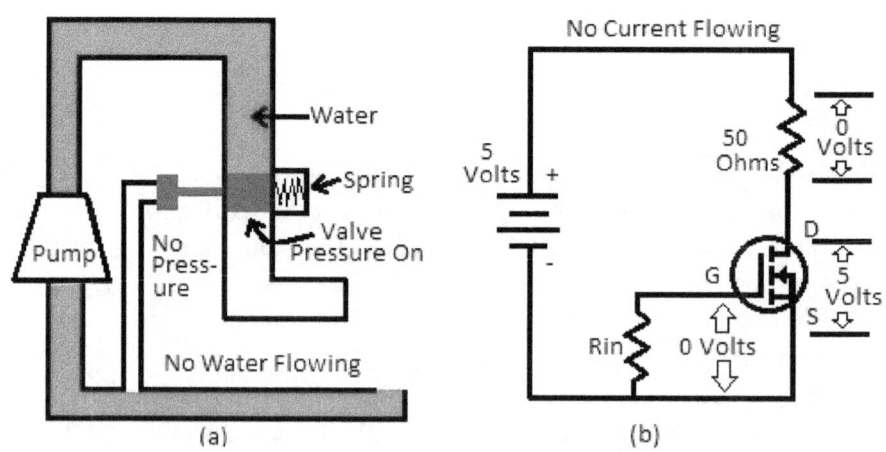

(a)   (b)

FIGURE 6-6

We find great similarities between Figure 6-6, part (a), the "off" drinking fountain, and part (b), the "off" MOSFET transistor circuit. Let's look at currents, first. No current flows anywhere. No water current, no electron current.

Let's look next at input pressures. Neither system has input pressure. In part (a), the on/off button gets no water pressure. The water coming from the bottom is not pressurized: it doesn't get pressurized until it comes out the top of the pump. In part (b), the gate has no voltage in relation to the source, because both paths lead to neutral, our circuit's 0 volts reference.

Now, let's look at output pressures. Both systems have output pressure. In part (a), the valve, not being turned on, is blocking the water flow. Hence, it is receiving the water pressure. In part (b), the drain-to-source, not being turned on, is blocking drain-to-source current flow. Hence, the drain-to-source is receiving the battery pressure.

Looking at Figures 6-5 and 6-6 one last time, we find the faucet and transistor systems almost identical. You merely need to substitute the words "water pressure" and "voltage", and substitute "water flow" and "current", to make the systems' descriptions sound the same.

To summarize, then, when we look at figures 6-2, 6-4, 6-5, and 6-6, we find that our MOSFET transistor circuit behaves like a controllable digital switch. We also find that it works much like the on/off water faucet that we studied earlier. Because of this similarity, we can replace our earlier faucet-based designs with transistor-based ones. Look at the picture below:

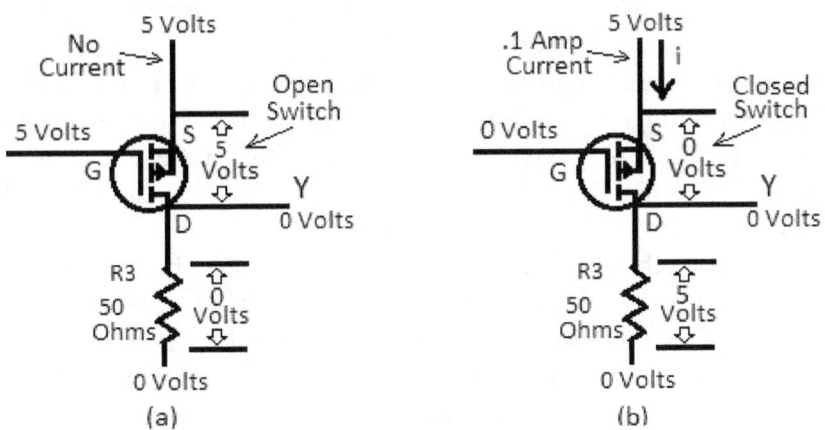

FIGURE 6-7

You might wonder where the battery is. Electrical circuits often don't show it, for a variety of reasons. The power supply voltage and neutral are usually drawn like this, as labels. In fact, for MOSFET circuits, 5 volts is often labelled "V_{DD}," and 0 volts is often labelled "V_{SS}."

Figure 6-7 is like our old Figure 4-6, for two faucets. In spite of all the complexity in understanding how electricity and transistors work, our circuit can really only do two things. It can only be in either state (a), on the left, or state (b), on the right.

Recall that we refer to the behavior in Figure 6-7 as an "inverter." +5 volts (gate-to-source) input gets inverted into 0 volts (drain-to-source) output in part (a). 0 volts (gate-to-source) input gets inverted into 5 volts (drain-to-source) output in part (b). In the picture below, I have replicated our earlier inverter truth table and logic symbol that we developed using a water faucet:

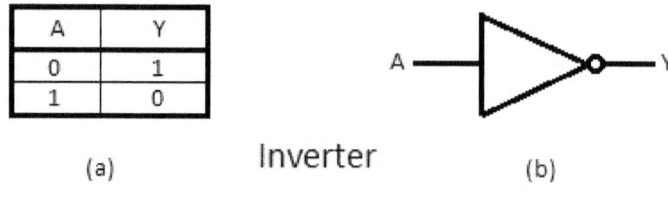

A	Y
0	1
1	0

(a) Inverter (b)

FIGURE 6-8

Notice how clear and uncluttered these two pictures are, compared to the actual circuit in Figure 6-7. The labels, A and Y, should help you understand how Figure 6-8 represents the actual circuit in Figure 6-7. Find the points marked A and Y in Figure 6-7. From the point of view of the logic, they are the only points that matter. That is why Figure 6-8 is easier to understand: everything else in the circuit has been thrown away to create Figure 6-8.

If you need a brief review, Figure 6-8, part (a) is the truth table for the inverter. "1" stands for "true," or "it is true that 5 volts is present." "0" stands for "false," or "it is false that 5 volts is present, so 0 volts must be present." A 1 is used instead of 5 because not all digital systems use 5 volts. Also, 1 is traditional and convenient in the formal study of logic and Boolean algebra. The first row represents everything you need to know about Figure 6-7, part (b). The second row represents everything you need to know about Figure 6-7, part (a). Figure 6-8 part (b) displays this information symbolically: it lets you know that A is the input to the inverter, and Y is the output. Together, these two items say, "0 volts in gives you 5 volts out," and "5 volts in gives you zero volts out."

Next, let's redraw Figure 4-11, and show how one transistor inverter can control a subsequent one. Look at the picture below:

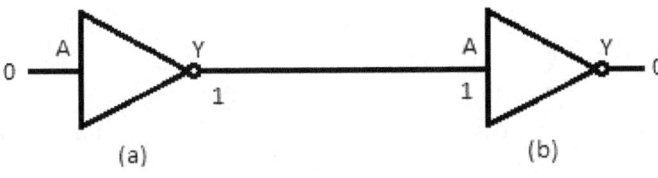

FIGURE 6-9

Do you remember this diagram? Let's see how to implement this logic using transistors instead of water valves. Look at the picture below:

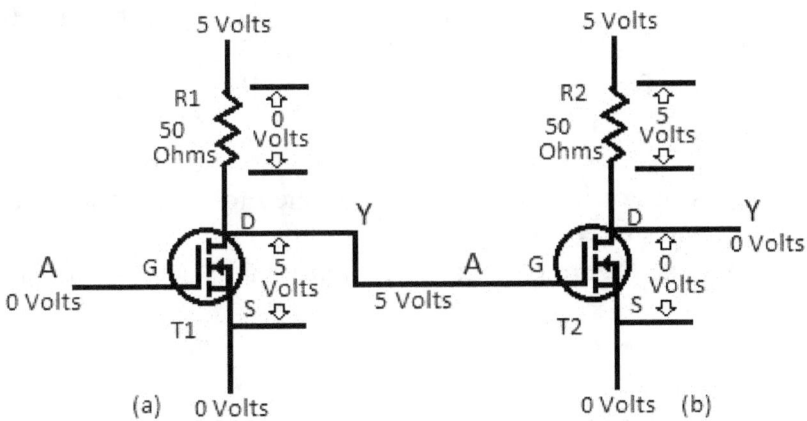

FIGURE 6-10

Figure 6-10 is the physical circuit symbolically represented in Figure 6-9. The Y-to-A line in the middle of Figure 6-10 is the connection between the first inverter and the second. Find where I have labelled the resistors R1 and R2, and the transistors T1 and T2. This labelling is standard practice, and saves words in the next paragraphs.

First, refer to Figure 4-4. Seeing Figure 6-10 in terms of water pressure will help you. Think of transistor T1 as a water valve that is turned off, since it has no pressure (voltage) at its input. Since pressurized current can't "drain" down through T1's drain, the pressurized current instead applies itself to the input "button" (gate) of T2. This turns on the output current flow down T2, drain-to-source, removing "pressure" (voltage) there.

Now let's speak purely in electronic rather than hydraulic, terms. Let's say in words what Figure 6-10 pictures. An external 0 volts, probably from some other logic gate, reaches the gate of the T1. This turns off current flow, drain-to-source, in T1. That becomes a current flow dead-end. You might think that current might instead flow across from Y to A and into the gate of T2. But, that's a current flow dead-end, too. By its high-resistance nature, no current ever enters the gate of the T2. That means no current going through the R1, no work is done to it, and so no voltage drops across it. Thus, the external 5 volts bypasses R1, and is applied directly to the gate of T2. This turns the drain-to-source output current of T2 on. That output current is pulled down from the 5 volt supply through R2. We designed our circuit so that all 5 volts of work/coulomb drops across R2. This leaves 0 volts remaining across T2's drain-to-source.

Or, to make a long story short, 0 volts enters the first inverter, and 5 volts comes out of it. That 5 volts is applied to the second inverter, and 0 volts comes out of it.

We can make Figure 6-10 easier to understand. Look at the picture below:

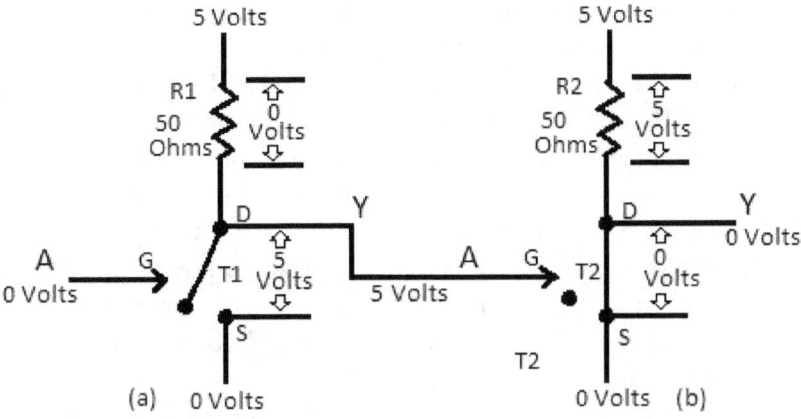

FIGURE 6-11

Figure 6-11 is the same circuit as Figure 6-10. But remember, we said transistors can act as switches, so let's replace T1 and T2 by switches. Since T1 is off, we replace it with an open switch, as seen in Figure 6-4. Since T2 is on, we replace it with a closed switch, a seen in Figure 6-2. Now, we have simplified from the complicated transistor to the simpler switch. But next, we can make this circuit simpler, still. Look at the picture below:

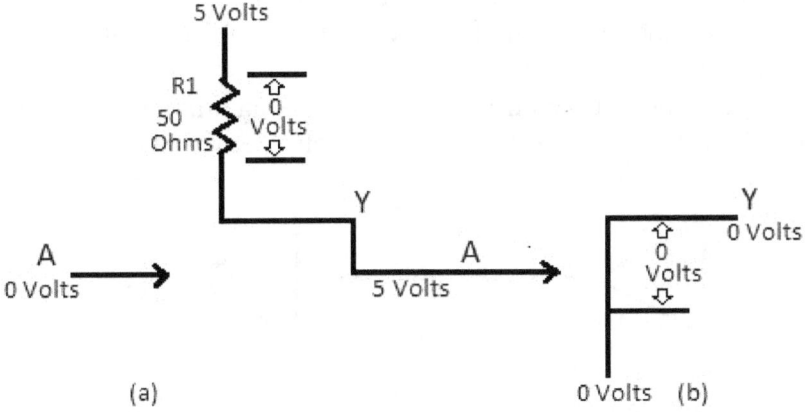

FIGURE 6-12

Now we have truly simplified the circuit. Figure 6-12 is the simplification of Figure 6-11, which itself simplified the actual circuit represented by the schematic in Figure 6-10. Look back at part (a) of Figure 6-11. T1's output is an open switch. An open switch is for all intents and purposes, nothing. It is air. No current goes through it. In a sense, it's not there. So, in Figure 6-12, part (a), I didn't draw it.

Next, look back at part (b) of Figure 6-11. T2's output is a closed switch. A closed switch is, for all intents and purposes, nothing but a wire. So, In Figure 6-12, part (b), I drew a wire leading to neutral. There was no need to draw R2 or the 5 volts feeding it, because they don't affect the output.

T1 and T2's gate inputs are super high resistances, so they don't allow input current (although they do affect output current). This is another reason not to draw T1 or T2.

Conceptually, this is how simple the circuit in Figure 6-10 truly is. Let's step through Figure 6-12. We see that 0 volts enters from the left, as A. With T1 theoretically gone, the 5 volts from the power supply comes in from the upper left, goes out of the left inverter's Y output, and enters the right inverter's A input. Finally, 0 volts comes out of the Y output of inverter 2, because no voltage ever appears across a wire (Remember? No work is needed to move electrons through no resistance.)

The trick to simplifying these circuits is to notice that the output side of each inverter divides into two halves: a switch and neutral on the bottom half, and a resistor and 5 volts on the top half. (See Figures 6-2 and 6-4.) But, if the switch is on, the output reduces to the wire and neutral on the bottom half. If the switch is off, the output reduces to the resistor and 5 volts at the top.

Now, let's look at the circuit in Figure 6-9 again, but this time with a logical 1, or 5 volts, entering on the far left. See the picture below:

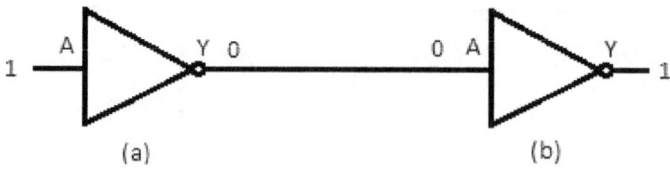

FIGURE 6-13

This time we will skip the steps from Figure 6-10--drawing the circuit--and Figure 6-11--replacing the transistors with switches. We will go directly to the step in Figure 6-12--drawing the simplified result. See that result in the picture below:

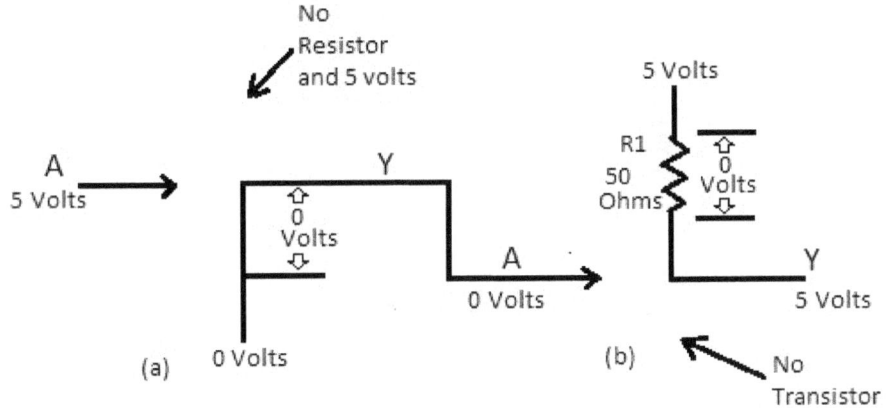

FIGURE 6-14

Figure 6-14 is the ultra-simplified version of what's inside Figure 6-13. We know that Figure 6-13 is an inverter controlling another inverter. We know that these inverters are MOSFET transistors. We follow the rules outlined above to simplify Figure 6-13 into Figure 6-14. 5 volts comes into the left inverter/transistor (the transistor switch is on), so we simply draw a wire to neutral. Neutral's 0 volts comes into the right inverter/transistor, keeping the transistor switch off; so we simply draw a resistor to 5 volts. That 5 volts exits the right inverter.

So, ultimately, the transistor circuit provides the next circuit with either a path to zero volts, or a path to 5 volts. We have pressure or no pressure; on or off; digital; binary.

Let's take another look at our OR gate, this time using transistors. As a quick reference, I have redrawn the truth table and logic diagram of the OR gate directly below:

A	B	Y
0	0	0
0	1	1
1	0	1
1	1	1

(a)

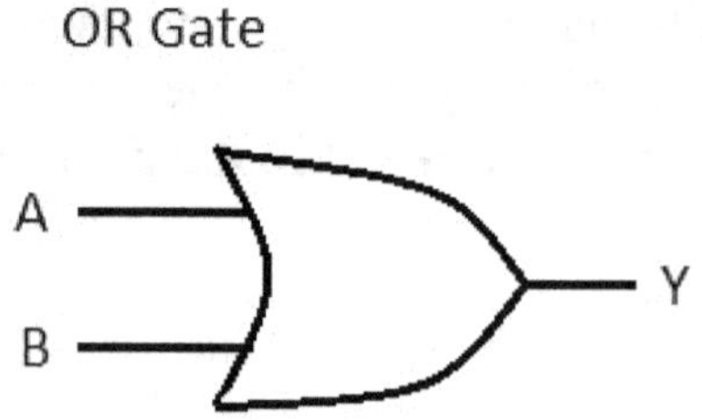

OR Gate

A

B

Y

(b)

FIGURE 6-15

Recall the concept underlying an OR gate. If one or more inputs are 1 (true, high), the output is 1. Now look at the transistor version of the OR gate, pictured below:

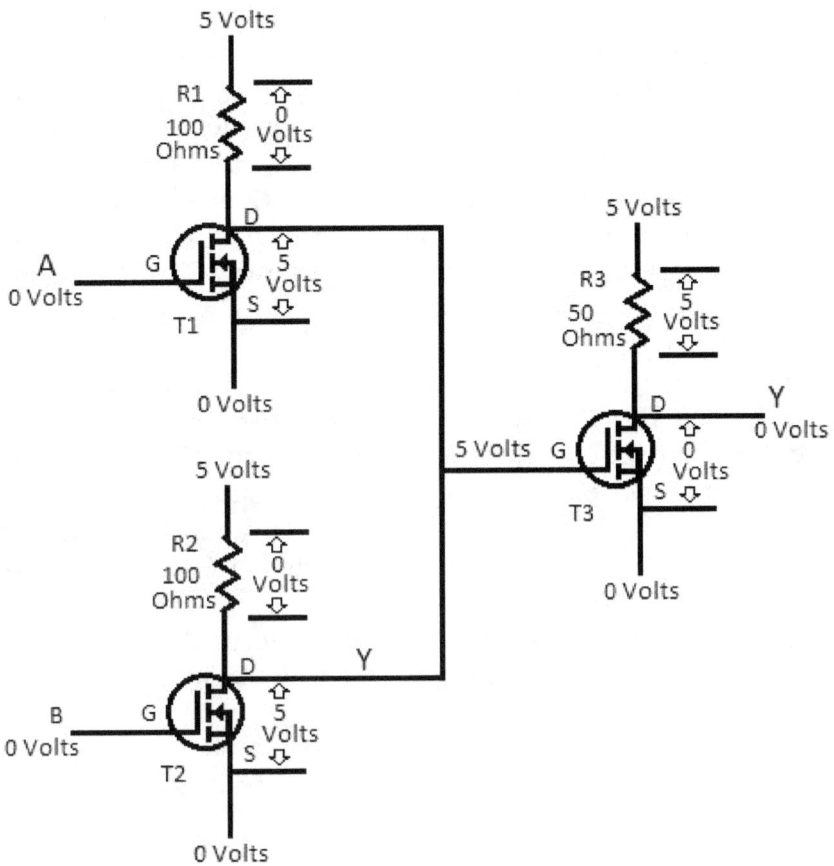

FIGURE 6-16

Figure 6-16 displays the first line of the truth table of the OR gate, as seen in Figure 6-15, part (a). When both the A and B inputs are low (zero volts), the Y output is low. This is the only condition in which the output is low. Refer back to Figure 4-12 to see the water fountain version of this circuit. It will help to give you a feel for the circuit flows and pressures.

In analyzing Figure 6-16, we find that both T1's and T2's output drain-to-source current flows are turned off. That's because both T1's and T2's input gate-to-source voltages are low. You should now cover

T1 and T2 with your thumbs: make them disappear. Since they are now behaving like open switches, they are effectively gone, from the point of view of their output sides. The only thing your eyes--and the input to T3--see are R1, R2, and 5 volts. Don't let two resistors (R1 and R2) and two 5 volts confuse you There is really only one 5 volts, labelled twice for convenience. And, R1 and R2 aren't doing anything at the moment, since no current flows through them. The 5 volts is applied (gate-to-source) to T3. It turns on turns T3's drain-to-source current, making it act like a closed switch. Since a closed switch has no voltage across it, 0 volts comes out of T3.

If you prefer to look at Figure 6-16 like a water faucet system, think of T1 and T2 as water valves that are turned off, since they have no pressure (voltage) at their inputs. Since pressurized current can't drain down through either T1 or T2, the pressurized current applies itself to the input "button" (gate) of T3. This turns on the output current flow down T3, drain-to-source, removing "pressure" (voltage) there.

Let's look at line number two of the OR gate, as shown in Figure 6-15, part (a). The picture below demonstrates line number two:

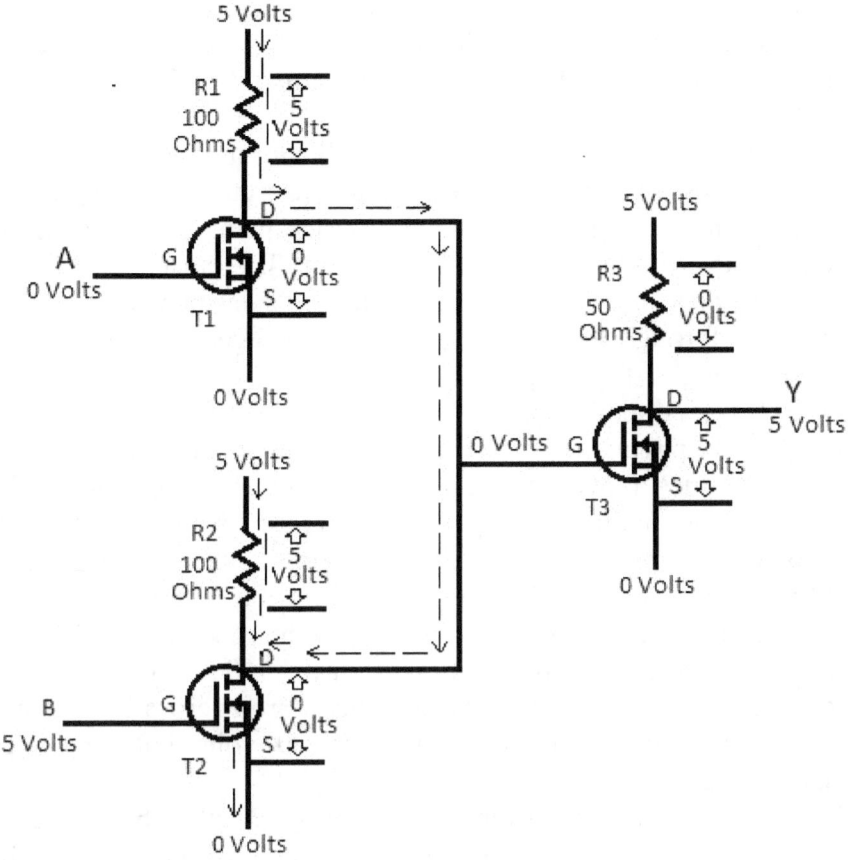

FIGURE 6-17

Summarily, if an OR gate has 0 volts input to A, and 5 volts input to B, 5 volts will exit Y. Transistor T2, in Figure 6-17, seems to behave normally. 5 volts input into its gate turns on T2's drain-to-source current, causing 5 volts to drop across R2, and no volts to drop across T2 (drain-to-source), the closed switch.

The behavior of transistor T1, however, is anything but normal. (It may help if you review Figure 4-14, the water faucet system, to understand this transistor system.) T1 has 0 volts coming into its gate. Hence, no current flows down T1's drain-to-source. When this happens, we expect R1 to have no voltage drop, and T1's drain-to-source have a five volt drop. But instead, we find both voltages are the opposite: R1 has a 5 volt drop, and T1's drain-to-source has a 0 volt drop. Why is this so?

The reason for the opposite voltages behavior around T1 is that the 5 volts above R1 has an alternate path to push current through. (See the dashed lines in Figure 6-17, to follow the paths of current flow.) Remember, even though T1 is off, T2 is on. The current from the 5 volt supply goes through R1, then through T2 instead of T1. Since current goes through R1, work is done, so voltage is dropped. Since the current avoids T1, it applies no pressure (voltage) to T1, so we measures 0 volts drain-to-source.

If you were paying very close attention, you noticed that I changed the resistor values above T1 and T2 from 50 ohms to 100 ohms. This is a bit advanced, but here's why I did that. The T2 we have been using pulls down .1 amps. In this case, it is pulling it through two, equal-valued resistors, R1 and R2. The .1 amp from the 5 volt supply splits evenly: .05 amps goes down R1, and .05 amps goes down R2. This makes the voltage drop across each 100 ohm resistor: $V=IxR=.05x100=5$ volts. The 5 volts voltage drop across Rout is now the same as in our inverter circuit, with 0 volts left across the transistors' drain-to-source.

Let's look at line number three of the OR gate truth table, as shown in Figure 6-15, part (a). The picture below demonstrates line number three:

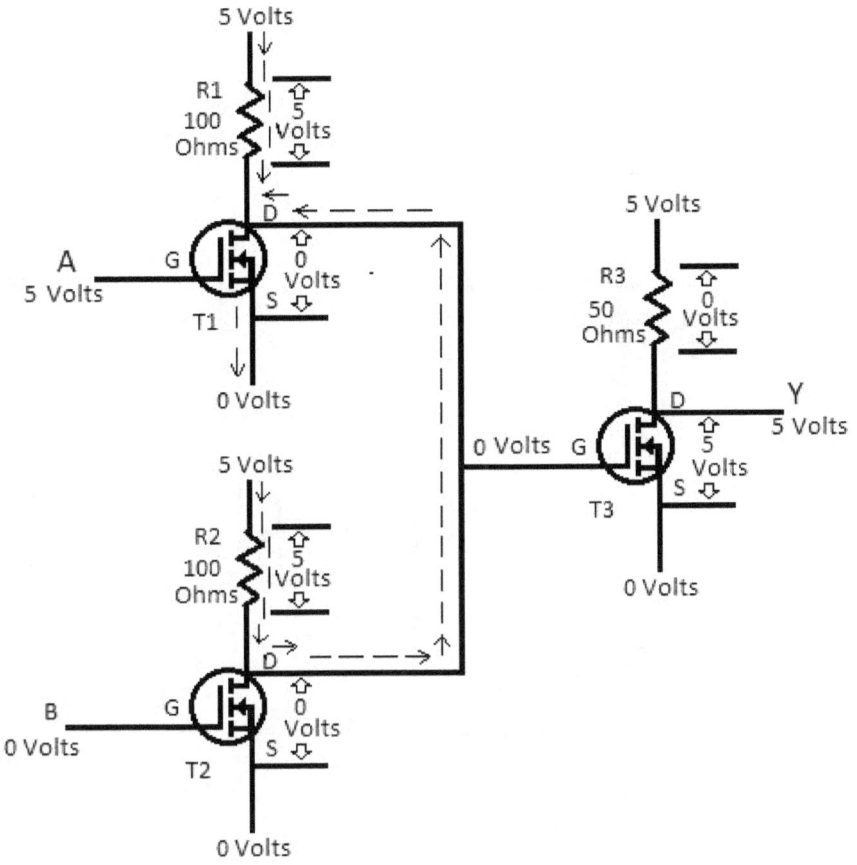

FIGURE 6-18

Summarily, if an OR gate has 0 volts input to B, and 5 volts input to A, 5 volts will exit Y. I won't spend much time on this circuit, since it behaves exactly like Figure 6-17: like line two of the truth table. The only difference is that the input voltages are reversed. The single 5 volts input is applied to A instead of B. The behaviors are the same. This time, though, T1 absorbs all the current flow coming through R1 and R2, since T1 is turned on. T2 absorbs none, since it is turned off. The result, 5 volts out of Y, is the same as in Figure 6-17.

Let's look at line number four of the OR gate truth table, as shown in Figure 6-15, part (a). The picture below demonstrates line number four:

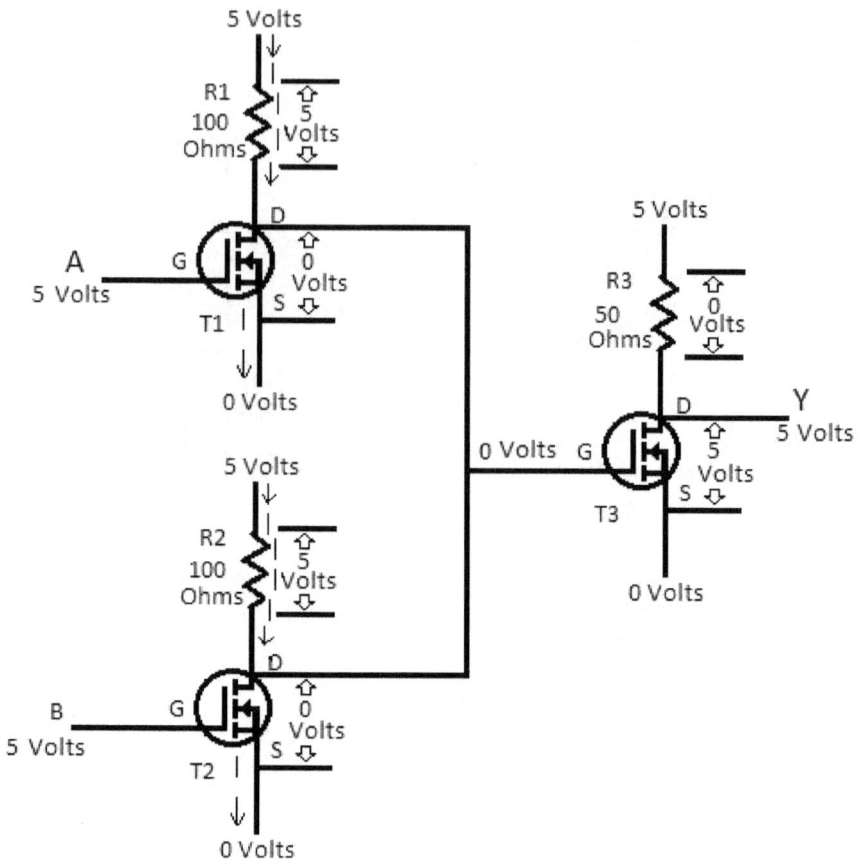

FIGURE 6-19

Summarily, if an OR gate has 5 volts input to B, and 5 volts input to A, 5 volts will exit Y. This circuit is actually easier to understand than the previous two in Figures 6-17and 6-18. Neither T1's nor T2's circuit voltages are "fooling us" into thinking they are on, when they are really off. T1 and T2 really are "on" this time, since they both have 5 volts applied to their inputs (gate-to-source). T1 and T2 each conduct current, behave like closed switches, or act like water faucets that are turned on (see Figure 4-15, if that helps you). But, like in Figures 6-17 and 6-18, the end result is the same. Since all current through R1 and

R2 is drained away from the gate of T3, no pressure (voltage) can be applied to T3. It stays off. Its output current stays off. Pressure (voltage) from 5 volts--through R3--builds up at T3, drain-to-source: at Y.

(Warning: you might want to skip this paragraph and the next two, because the topic is a bit advanced.) The change of R1 and R2 to 100 ohms solved a problem for us when only R1 or only R2 conducted current. But, have we just introduced a new problem? Let's do the math. If T1 pulls down .1 amp and R1 is 100 ohms, then Ohms Law says V=IxR=.1x100=10 volts of work is done getting current through R1. We could say exactly the same thing about T2 and R2. Doesn't that break the Law of Conservation of Energy? Aren't we accomplishing more work (10 volts per coulomb) than energy supplied (5 volts per coulomb)?

We can't break the laws of nature. The Law of Conservation of Energy, and the transistor itself, prevents this from happening. The transistor will self-limit its current flow. The transistor will cut back on current flow, down to the point where 0 volts drops across it (drain-to-source) and all the 5 volts of supply voltage drops across the resistor. Thus, rather than .1 amps, T1 will draw .05 amps through R1, then into itself. This will create V=IxR=.05x100=5 volts work done across R1, and none done across T1. T2 will do the same.

Because of the above characteristic of transistors, we usually choose a much higher value of drain resistor than the 50 or 100 ohms we have been using. As the resistor gets higher, the transistor's drain-to-source current gets lower. Less current means less power consumption. This means less heat. When we are putting millions of these transistor circuits in a microprocessor, we must minimize heat.

We have covered three of the fundamental logic gates: the inverter (Figure 4-10), the OR gate (Figure 4-16), and the NOR gate (Figure 4-21). We need to cover just a few more. But, we are going to take a short-cut. Remember that an inverter, in its simplest form, is just a one-transistor, one resistor circuit. Pictorially, we can say:

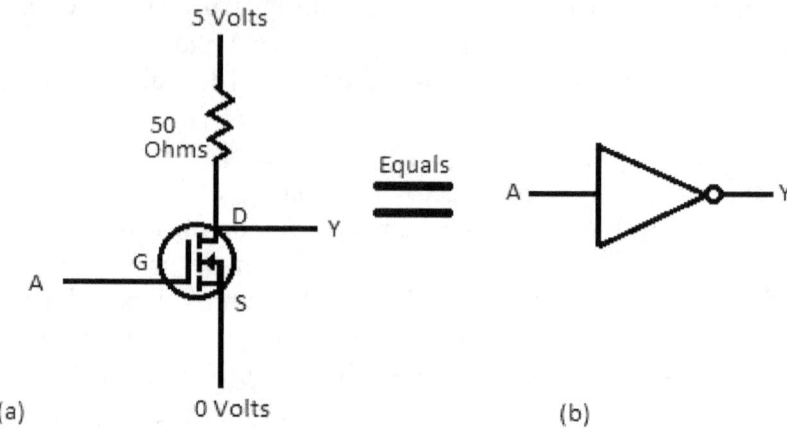

FIGURE 6-20

Figure 6-20 implies that, if we want to, we can draw the simpler inverter logic diagram in part (b), in place of the more complicated circuit diagram in part (a). With this short-cut, we realize that we can draw the OR gate circuit from Figures 6-16, 6-17, 6-18, and 6-19 in the following, more compact, form in part (a):

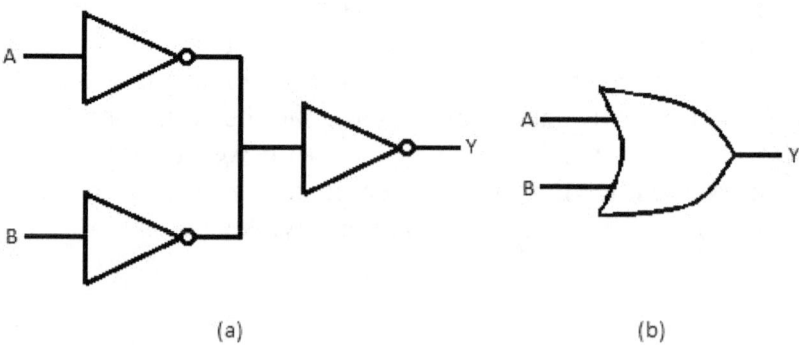

FIGURE 6-21

To draw Figure 6-21, part (a), all I did was replace each transistor-resistor pair from Figures 6-16, 6-17, 6-18, and 6-19 with the inverter symbol. Now, Figure 6-21, part (a), is certainly much simpler than those we encountered in Figures 6-16, 6-17, 6-18, and 6-19. But it is not simpler than the logic symbol for the OR gate in Figure 6-21, part (b). So, why did I bother to do this?

I did it so that you could see an example of our approach for the remaining logic gates. Rather than go through the lengthy circuits drawings and circuit analyses, we will simply design the remaining logic gates with inverters. After all, digital electronics and microprocessors are pretty much all transistors. If you can just get comfortable thinking of each inverter as a transistor/resistor circuit, you will have no loss in understanding.

The next logic gate we want to study is the AND gate. The truth table and the logic symbol for the AND gate are in the picture below:

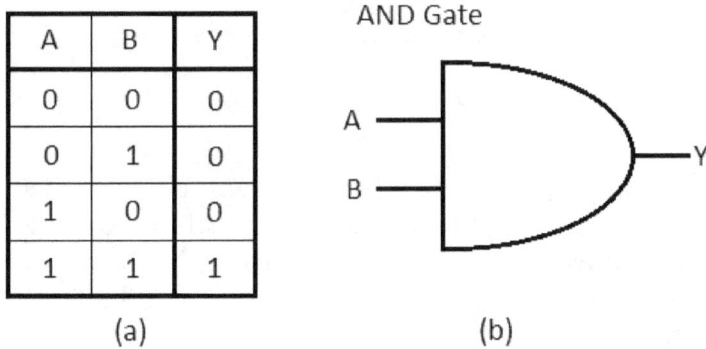

A	B	Y
0	0	0
0	1	0
1	0	0
1	1	1

(a)

(b)

FIGURE 6-22

Remember, each one of the four rows of the truth table in part (a) is a possible state of the AND gate, drawn in part (b). In each row, A and B represent a pair of input states; Y represents the output state for those inputs. It is called an AND gate because only when A **and** B are both 1 (true, high, 5 volts) will the output Y be 1. Find that state in the fourth row of the truth table. Any other pair of inputs produces a 0 (false, low, 0 volts) output.

There are multiple ways of making logic gates from transistors. If we want to, we can make an AND gate the following way:

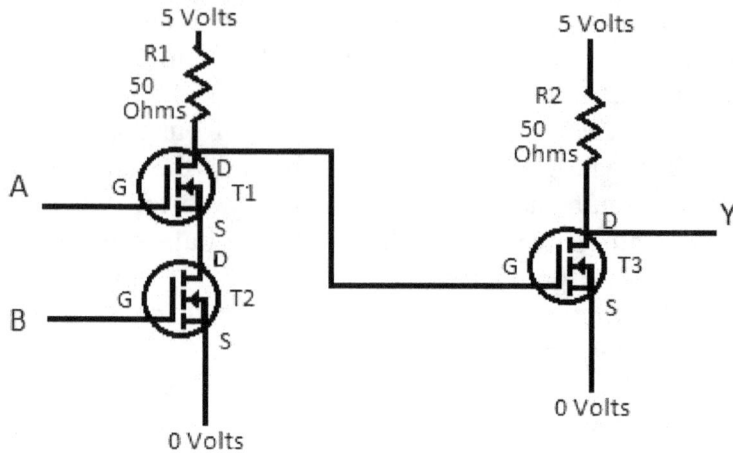

FIGURE 6-23

In Figure 6-23, only when both T1 *and* T2 are turned on (by 5 volts at A *and* B) will current flow through R1, T1 and T2. Only then will T3 be turned off, producing 5 volts at Y. This would be like having one water faucet with two valves and two buttons you would have to press to turn on the water.

Let's not go that route. Even though this is a more efficient design, let's just continue designing all of our gates with our same fundamental building block, the transistor/resistor inverter of Figure 6-20. Our goal here is not for you to become a designer of optimally efficient logic gates. Our goal is for you to understand how microprocessors work. We want minimal confusion, and I am afraid that designing different circuits for each gate will confuse matters. Plus, I find beauty in the simplicity of building our mighty microprocessor from one simple circuit.

Below is a picture of how to create an AND gate, using our short-cut method of drawing only inverters. Remember: think of each inverter as our transistor/resistor circuit of Figure 6-20, part (a).

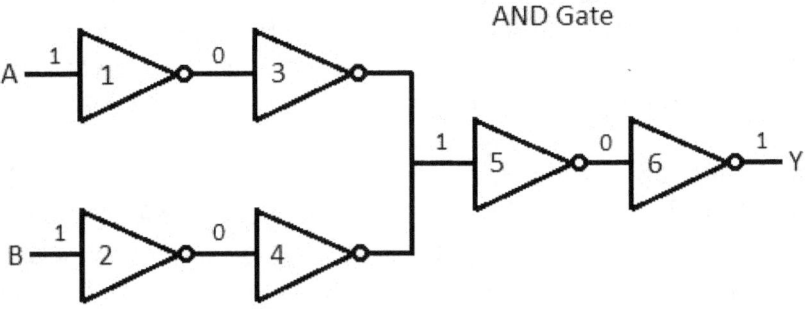

FIGURE 6-24

In Figure 6-24, I have labeled the six transistor/resistor circuits (inverters) with numbers inside the inverter triangles. Outside the triangles, I have labeled the logic states for row four of the AND gate truth table: logic 1's (5 volts) on the A and B inputs producing a logic 1 on the Y output. This one is pretty easy to figure out, since each inverter takes what's on the input and produces the opposite on the output.

For the AND circuit, I will only show one more line of the truth table: line three. See the picture below:

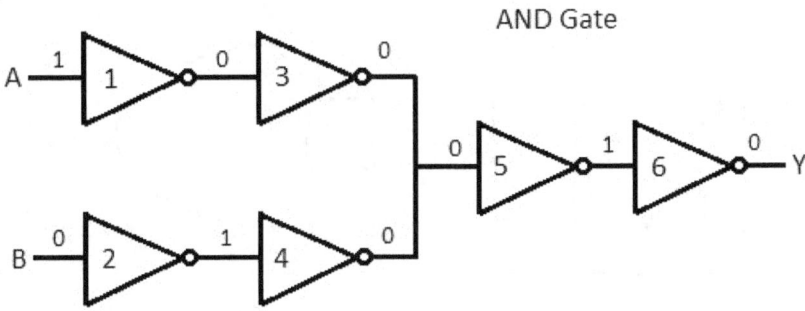

FIGURE 6-25

Line three of the truth table for an AND gate says that a 1 (5 volts) on the A input and a 0 (0 volts) on the B input produces a 0 on the Y output. You should notice something funny about inverter number 3: logic 0 (0 volts) on its input produces logic 0 on its output. It doesn't invert. And yet inverter number four does produce the correct output: a logical 1 (5 volts) on its input produces a logical 0 on its output. It does invert.

It should dawn on you that they can't both work correctly. If inverter 3 produced a 1 at its output and inverter 4 produced a 0 at its output, you would have the input to inverter 5 be simultaneously 1 and 0. Logically, it would be like something being true and false at the same time. Voltage-wise, it would be like having both 5 volts and 0 volts at the same place.

So, why does inverter 4 win the logic battle with inverter 3, so that it produces the correct output, forcing inverter 3 to accept the (for it) wrong output? You already know the answer (The off transistor in inverter 3 can't build up pressure, because current bypasses it and heads to the on transistor in inverter 4.) Refer back to Figure 6-17, if you don't remember. Inverters 3, 4, and 5 of Figure 6-25 are exactly the same as the ones pictured in Figure 6-17. Review that material, if you need to. Inverters 1, 2, and 6, are just transistor/resistor pairs added before and after that circuit, which you already understand.

The next logic gate we want to study is the NAND gate. The truth table and the logic symbol for the NAND gate are in the picture below:

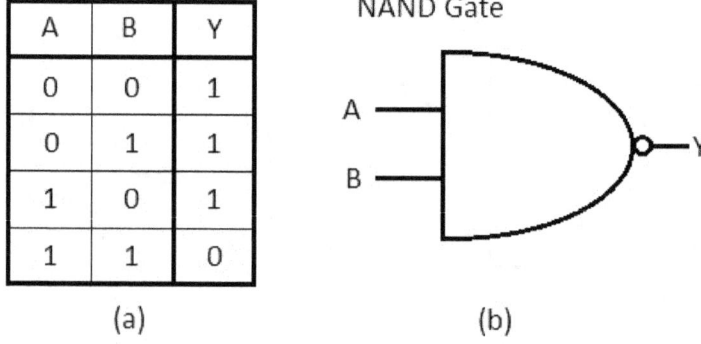

A	B	Y
0	0	1
0	1	1
1	0	1
1	1	0

(a)   (b)

FIGURE 6-26

The name NAND gate is short for Not-AND gate. Compare the truth table for the NAND in part (a), above, with the truth table for the AND in Figure 6-22, part (a). You will find that the NAND is like the AND with each row's Y output inverted. That's why the logic symbol for the NAND, in part (b), above, looks just like the logic symbol for the AND, but with a bubble affixed to the Y output. The bubble symbolizes that the output is inverted. Put into words, a NAND gate produces a 0 output (false, low, 0 volts) only when both the A *and* B inputs are high.

Look at the simplified, inverter-style diagram for the NAND gate, below:

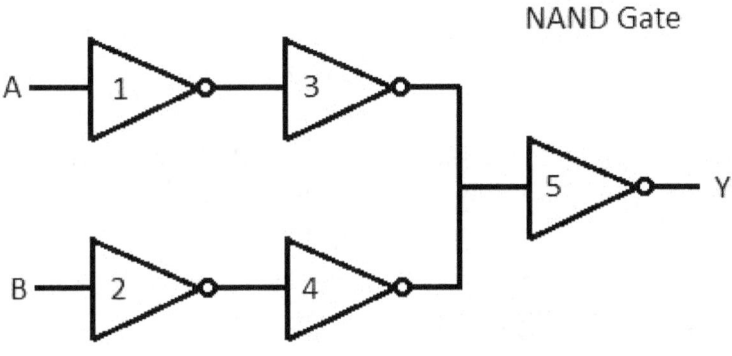

FIGURE 6-27

Perhaps you remembered the way we turned an OR gate into a NOR gate, way back in Figure 4-18. We added an inverter to the output of an OR. Rather than adding a seventh inverter to the output of the AND gate, in Figure 6-24, we simply removed the last one. It gives the same affect.

Using this same trick of removing an inverter instead of adding one, let's revisit Figure 6-21. That is an inverter-style diagram of an OR gate. Rather than adding an inverter to the output of the last inverter, to create a NOR gate, we can remove the last inverter, instead. We still create a NOR gate, and it looks very simple, as you can see in the picture below:

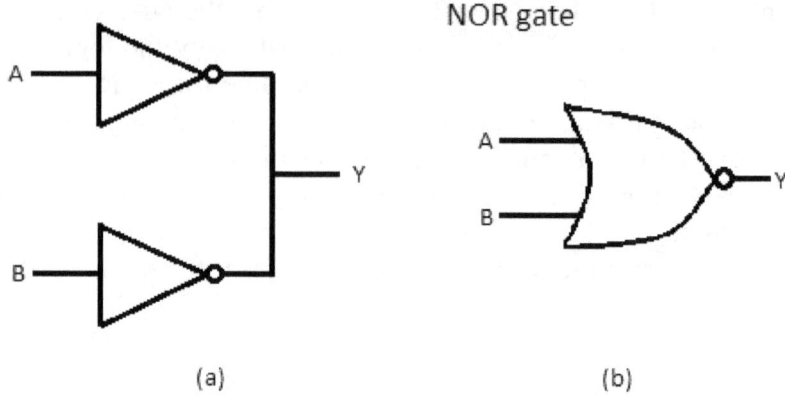

NOR gate

(a)   (b)

FIGURE 6-28

The next logic gate we shall study is the XOR gate. The truth table and the logic symbol for the XOR gate are in the picture below:

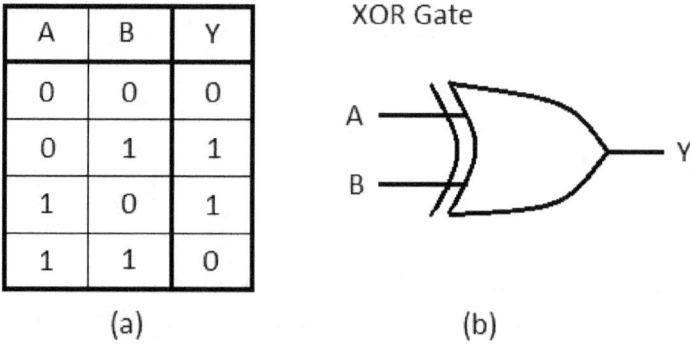

A	B	Y
0	0	0
0	1	1
1	0	1
1	1	0

XOR Gate

(a)   (b)

FIGURE 6-29

One way to describe this XOR gate is that its truth table looks just like the truth table for an OR gate with one exception: the fourth row's output is 0. Another way is to say that a 1 into A *or* B, but *not* both, produces a 1 out of Y. But, the best way to describe the truth table involves looking at each row separately. For each row, count the number of 1's on the inputs. If you count an odd number of 1's, then the Y output will be a 1. If you count an even number of 1's, then the Y output will be a 0. No 1's is an even number of 1's, so row number one will have a 0 output on Y. Row number two has only one 1, in the B column. One is an odd number, so the Y output of row two is 1.

The reason we use this more elaborate way of describing the behavior of the XOR gate involves something I have not yet explained: gates (except for inverters) are not limited to two inputs. You can have three or more inputs. However, for the AND, OR, NAND, and NOR gates, extra inputs are not a "big deal." Their behaviors are a logical extension of two-input gates.

Briefly, for three inputs, A, B, and C, we have eight possible rows of a truth table. A three-input AND gate behaves like a two-input AND gate: only the last row has a Y output of 1 (Figure 6-30, part (b)); a 1 is produced only when A, B, *and* C inputs are 1's. A three-input NAND gate behaves like a two-input NAND gate: only the last row has a Y output of 0; a 0 is produced only when A, B, *and* C inputs are 1's. A three-input OR gate behaves like a two-input OR gate: only the first row has a Y output of 0 (Figure 6-30, part (a)); a 1 is produced when A, B, *or* C inputs are 1's. A three-input NOR gate behaves like a two-input NOR gate: only the first row has a Y output of 1; a 0 is produced when A, B, *or* C inputs are 1's. Look at the picture below to see what I mean:

OR Gate

A	B	C	Y
0	0	0	0
0	0	1	1
0	1	0	1
0	1	1	1
1	0	0	1
1	0	1	1
1	1	0	1
1	1	1	1

AND Gate

A	B	C	Y
0	0	0	0
0	0	1	0
0	1	0	0
0	1	1	0
1	0	0	0
1	0	1	0
1	1	0	0
1	1	1	1

(a)   (b)

FIGURE 6-30

The behavior of XOR gates as we move from two inputs to three or more inputs, is not so easily summarized. Look at the picture of the three-input XOR gate in part (a), below:

XOR Gate   XNOR Gate

A	B	C	Y
0	0	0	0
0	0	1	1
0	1	0	1
0	1	1	0
1	0	0	1
1	0	1	0
1	1	0	0
1	1	1	1

A	B	C	Y
0	0	0	1
0	0	1	0
0	1	0	0
0	1	1	1
1	0	0	0
1	0	1	1
1	1	0	1
1	1	1	0

(a)   (b)

FIGURE 6-31

Looking at the first four rows of Figure 6-31, part (a), you recognize the same Y outputs as the four rows of Figure 6-28. But, the Y outputs of the next four lines of Figure 6-31 are the reverse of the Y outputs of the first four lines of Figure 6-31. Because of this "strange" behavior, we use the more complicated description of the XOR gate behavior. An odd number of 1's on the inputs produces a 1 on the output. An even number of 1's on the inputs produces a 0 on the output.

A three-input version of a gate is easy to draw. See the picture below:

3-Input OR Gate

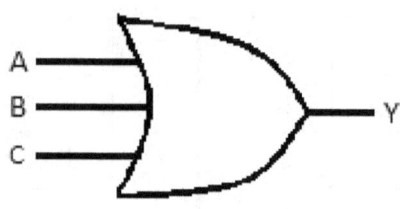

FIGURE 6-32

Internally, to add the third input (C), we simply add a new transistor/resistor inverter behind the C input line, and attach it internally the same way the A and B inverters are attached.

The last logic gate we shall study is the XNOR gate. The truth table and the logic symbol for the 2-input XNOR gate are in the picture below:

A	B	Y
0	0	1
0	1	0
1	0	0
1	1	1

(a)

XNOR Gate

(b)

FIGURE 6-33

As you might predict, the XNOR gate is the XOR gate with its output inverted, for each line of their corresponding truth tables. The bubble on the output of the logic symbol also indicates this. We describe the behavior of the XNOR gate by saying: an odd number of 1's on the inputs produces a 0 on the output. An even number of 1's on the inputs produces a 1 on the output. Refer to the three-input XNOR truth table, in Figure 6-31, part (b), to see why this description is necessary, and to prove to yourself that it works.

Let's look at the circuit that creates the XOR gate. Look at the picture below:

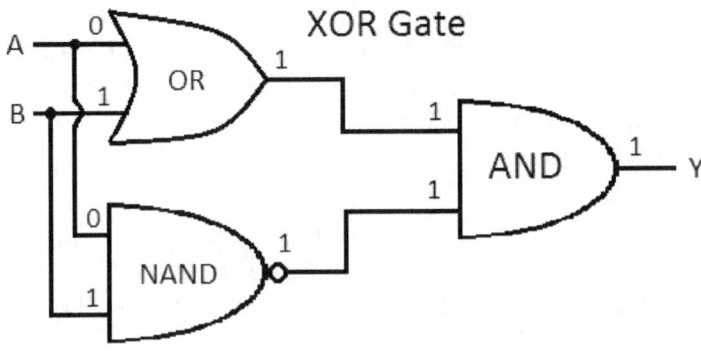

FIGURE 6-34

At first glance, Figure 6-34 might disturb you. You expect to see a circuit for the XOR gate, and instead you see three other gates, connected together. But remember: all of these figures are pictorial or symbolic ways of explaining how our basic transistor/resistor inverter circuits are connected together. We already know all we really need to know about the behavior of the XOR gate from the truth table in Figure 6-29.

So, we learn how the XOR gate works by navigating through several levels of diagrams. We start with the logic symbol in Figure 6-29, part (b). We learn that the circuit inside this behaves like the three interconnected gates in Figure 6-34. We remember that the NAND gate in Figure 6-34 behaves like the five interconnected inverters in

Figure 6-26; the AND gate behaves like the six interconnected inverters in Figure 6-24; and the OR gate behaves like the three interconnected inverters in Figure 6-21. Finally, we remember that the inverter circuit repeated fourteen times--but connected in very specific ways described by these diagrams--is simply the actual circuit described by Figure 6-20.

This is an important moment in your understanding of microprocessors. You have just found that, from the basic building block of the transistor/resistor inverter, we build the building blocks called gates. From these building blocks, we built a more complex building block, the XOR. We will keep building more complex building blocks from simpler blocks, until we have a whole microprocessor.

One last point concerning Figure 6-34: notice the 0's and 1's I have placed on the inputs and outputs of the gates. I have set up for you a demonstration of line number two of the truth table for an XOR gate. We know that a 0 on the A input and a 1 on the B input produces a 1 on the Y output. Part of the fun of digital electronics is watching the flow of the logic as it ripples through the gates. Test your knowledge of the truth tables for the gates that make up this circuit. Make sure you understand why each 1 or 0 appears where it does. Try solving for the other three lines of the XOR truth table.

CHAPTER SEVEN
LATCHES

The logic gate is one of the two basic building blocks of digital circuitry. The other basic building block is the latch. Just as there are several kinds of logic gates--AND, OR, etc.--there are several types of latches--D, RS, clocked RS, JK, etc. The latch also has a sibling, the flip-flop, which we will encounter later.

What is a latch? A latch is the fundamental unit of storage in a logic circuit. A latch stores one bit.

What do we mean by storage? When we delve deeper into microprocessors, we will find that much of what goes on involves moving information around. A group of bits moves from the keyboard into the microprocessor; from the microprocessor to the video display; from the hard drive to the system memory, etc. Information moves from point A to point B.

The point is, once it arrives at point B, it must "stick" there. It must stay there until we choose to replace it with new information, a new group of bits; those, too, must "stick" there at point B. Furthermore, when we move information from point A to point B, the information does not disappear from point A. It "sticks" there, too, until we choose to change it. Thus, when we "move" a group of bits from point A to point B, we are making an exact copy at point B of the bits that are at point A. This temporary replication and retention of information from the A group of transistors to the B group of transistors is "storage."

Why do we need storage? Microprocessors are sequential devices. Things happen in an orderly process: step 1 happens, followed by step 2, etc. Furthermore, step 2 will often operate on the information involved in step 1. It can't operate on it if it is gone. It can only operate on it if it (step 1) is stored.

For example, microprocessors can add two 8-bit (8-transistor) numbers together. This is a sequential operation: step A happens at time 1; step B happens later, at time 2; and step C happens later, at time 3. If step 1 puts number A in storage area A, and step 2 puts number B in storage area B, step 3 needs group A and group B to still be there when it adds A to B.

There's another reason microprocessors need storage. The first reason, above, had to do with things happening sequentially. The second reason, below, has to do with avoiding things happening simultaneously. For reasons of efficiency, microprocessor systems often have one device connected to many devices. See part (a) in the picture below:

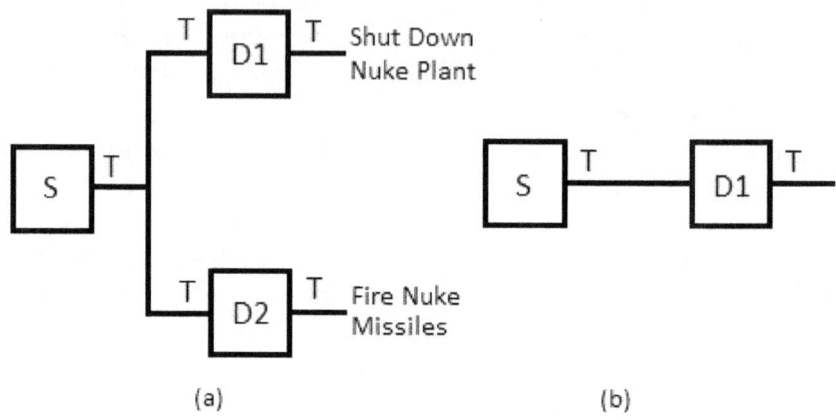

(a)   (b)

FIGURE 7-1

Part (a), above describes a source device, labelled S, connected to two destination devices, labelled D1 and D2. Imagine you are S. You are binary. You communicate one command--either T for true or F for false--by holding up a sign. One side of the sign is marked, T; the other side is marked, F.

D1 and D2 are your two co-workers. They are also binary. D1 responds to the T sign by shutting down the nuclear plant; he responds to the F sign by not shutting down the nuclear plant. D2 responds to the T sign by firing the nuclear missiles; she responds to the F sign by not firing the nuclear missiles.

Now, imagine in Figure 7-1, part (a), that S finds out about a failure at the nuclear power plant, and must tell that to D1, so D1 can shut down the nuclear power plant. As shown above, S displays the T side of the sign. D1 shuts down the nuclear power plant. Unfortunately, D2 also sees the T sign. Since she has no way of knowing the information is not intended for her, she fires the nuclear missiles.

Armageddon follows.

So, we need temporary storage devices at D1 and D2. When we need to shut down the nuclear power plant, we need to be able to open up a door into D1--but not into D2--and store a T command sign with D1, until it is safe for him to come out of shutdown. Since we did not open up the door into D2, she still retains an F command sign, so she does not fire the nuclear missiles.

Now, look at Figure 7-1, part (b). In a simple one-to-one system like this, S could just sit there as long as necessary with the T information, because it has nothing else to do. D1 does not have to be a storage device; D1 can be a gate. But, looking back at Figure 7-1, part (a), S does have something else to do. S has to communicate with D2, separately from D1. So D1 and D2 must be storage devices: latches; they can't be gates.

Furthermore, D1 and D2 must be more than just storage devices: they must be *selectable* storage devices. I already mentioned "opening a door" before storing information. If D1 and D2 were as simple as shown in Figure 7-1, part (a), we would wind up with the same problem we started with: when we store a T at D1, a T would also be stored at D2, and the missiles would incorrectly be fired. So, the physical design of our latch circuit must incorporate the ability to be selected: the ability to have its door opened. Look at the picture below:

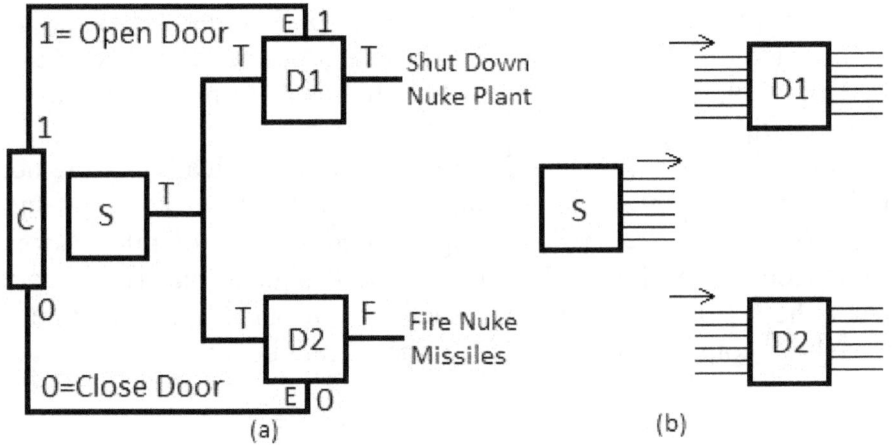

(a)   (b)

FIGURE 7-2

Figure 7-2, part (a), shows the behavior we want--the nuclear plant shut down, and the nuclear missiles not fired. To accomplish this, we must add some additional control circuitry, external to our storage circuitry. Our control circuitry, labelled C, is the "door opening" *sending* circuitry. Whenever S has a 1 or 0 to store in either D1 or D2, C must send a 1 to the "door opening" *receiving* circuitry, labelled E, built into the desired latch. If C sends a 0 to either D1, D2, or both, the output of S is ignored by D1, D2, or both, respectively.

Let's trace through how this works in Figure 7-2, part (a). S sends out the T (true) sign. However, since the control circuit, C, has sent a 0 to D2's E input, D2 can't receive the T; the 0 on E closes D2's door. Notice that, even though D2 has a T at its entrance (input), the T is not at its exit (output).

On the other hand, D1 does receive the T, because its door is open. Notice that D1 has a T at its entrance (input), and the T passes through to its exit (output). It passes through because the control circuit, C, has sent a 1 to D1's "door opener." The 1 on its E opens D1's door.

The outputs of D1 and D2 will stay at the logic level--T or F--that was present at their inputs the last time the last time their door was opened with a logic 1 on their E inputs. D1 and D2 ignore the state of their inputs when the door opener is closed with a logic 0.

You may wonder why we go through all this effort. I claimed this is done for greater efficiency. But, it may look like less efficiency to you, because we have added more wire paths, not less. And, we have added new control circuitry, as well as the need for more complexity inside our flip-flops.

However, look at Figure 7-2, part (b), above. Typically, we are not sending one bit of information from S to its destinations, as in part (a). In part (b) I have drawn S needing to send six bits of information, simultaneously, to either D1 or D2. If the design of part (b) did not include the select-and-store design elements of part (a), S would need twelve lines coming out of it. Look at the picture below:

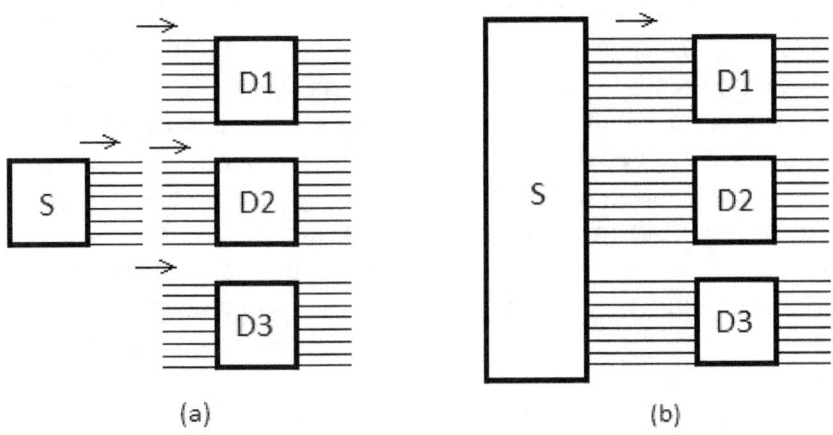

(a)   (b)

FIGURE 7-3

In Figure 7-3, part (a), we have used the select-and-store tricks of
Figure 7-2, part (a). (We haven't drawn that extra control circuitry, to
prevent confusion.) Using these techniques, we are able to transfer and
store eight bits of information--one byte, by definition--to either D1,
D2 or D3. We have kept the number of lines coming out of S to just
eight. A line is a copper wire, or a copper trace on a circuit board.
Perhaps it is a pin or lead sticking out of an integrated circuit chip (IC).

Now look at Figure 7-3, part (b). Here, we have not used the
select-and store tricks of Figure 7-2, part (a). Now, S must have
twenty-four separate lines. Those lines are totally dedicated to
maintaining the outputs of D1, D2, and D3. They can never be used for
anything else.

Now, imagine a real microprocessor system. S might be required to
send to ten devices: D1 through D10. Instead of sending eight bits to
D1 through D10, S might send sixty-four bits. Since 64x10=640, S in
Figure 7-3, part (b) would have 640 lines coming out of it. This would
create an impossibly huge microprocessor. Using Figure 7-3, part (a),
S would have sixty-four data bit connections: a much more
manageable number.

Before we move on to describing how we will build our latch, let's
take a moment to appreciate the wonder and strangeness of what we
are describing. We are talking about information moving around, on its
own, from one tiny place to another. It's like you writing a message,

making a copy of it, walking the copy to a friend's house, and handing it to the friend. But, in the microprocessor, there is no "you." It all happens automatically, under machine control.

We are talking about creating a pattern or sequence of voltages (5 volts) and non-voltages (0 volts), in transistors. This whole concept of voltages is hard enough to comprehend. Voltage is "pressure-like." We have one battery, but many voltages. Then, amazingly, we move this pattern somewhere else, to a different group of transistors. The voltages were in one place, and now they are in two places.

Now, it is time to design our latch. Of course, there are different ways to build a latch, but I enjoy the elegance and uniformity of building our system from the same building block, the transistor-resistor inverter seen in Figure 6-20. Look at the picture below:

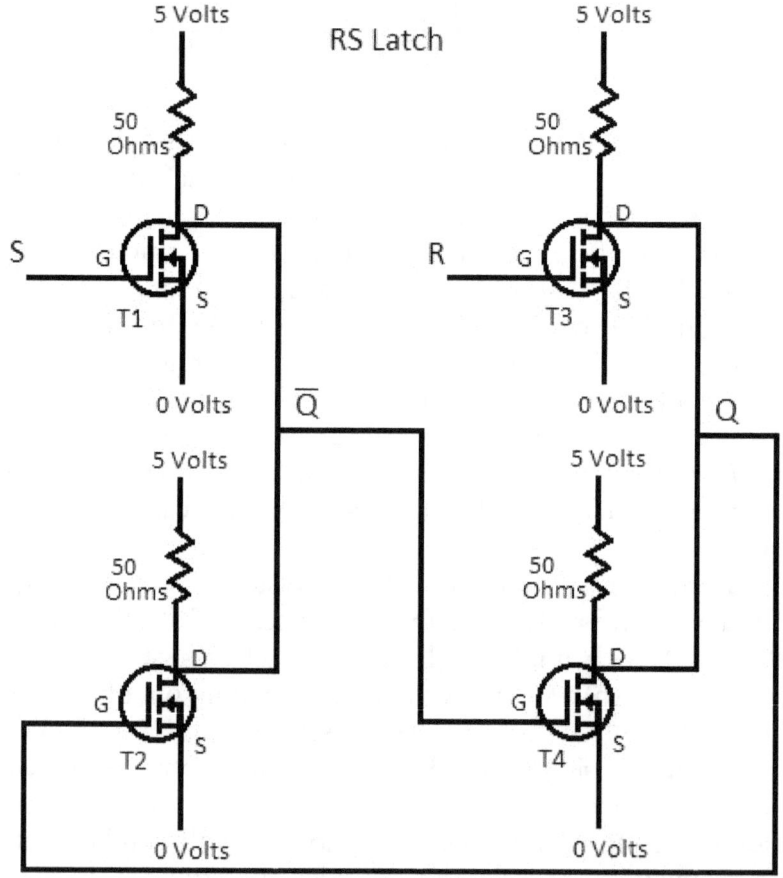

FIGURE 7-4

Figure 7-4 is the first step in building our latch. The first thing to notice is our basic transistor-resistor inverter, repeated four times, and labelled T1 through T4.

Next, notice how the drains of T1 and T2 are connected together. We have seen this arrangement before. If you look back at Figure 6-28--two inverters with their outputs connected making a NOR gate--and realize from Figure 6-20 that an inverter is our basic transistor-resistor circuit, you will realize that T1 and T2 have been connected to create a NOR gate. Since T3 and T4 are connected the same way, drain to drain, they too have been connected to create a second NOR gate.

Third, notice how the output of the T1-T2 NOR gate connects to the gate input of T4. Symmetrically, the output of the T3-T4 NOR gate connects to the gate input of T2.

Fourth, notice that the input gate of T1 is labelled S, and the input gate to T3 is labelled R. These are our Set and Reset inputs, our means to control the logic state of the latch.

Fifth, the output of NOR gate T1-T2 is labelled Q with a line above it, or "NOT Q." The output of NOR gate T3-T4 is labelled Q. Q is our stored output bit. NOT Q is the opposite of Q, meaning if Q is 1 (5 volts), NOT Q is 0 (0 volts). If Q is 0 (0 volts), NOT Q is 1 (5 volts). We will soon ignore NOT Q. It comes in handy for individual, stand-alone latches in digital electronics, but microprocessors usually involve only the Q outputs of parallel collections of latches.

Look at the picture below, to see where we are so far:

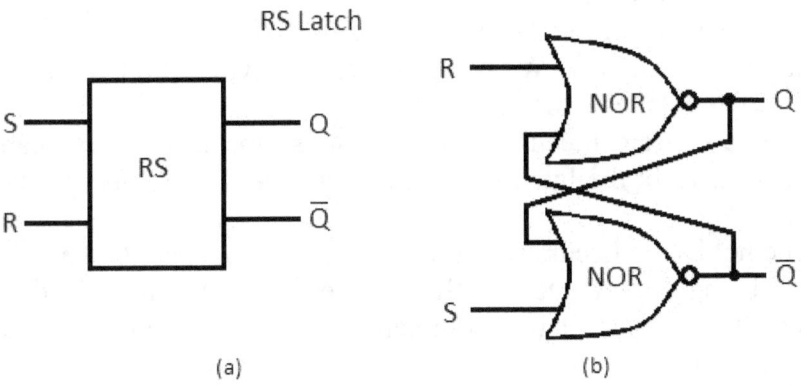

(a)   (b)

FIGURE 7-5

Part (a), in Figure 7-5 is the rather dull logic symbol for the RS latch. It's very high-level: it only shows the two inputs, set and reset, on the left; and the two outputs, Q and NOT Q, on the right. Part (b), in Figure 7-5, is a much neater and tidier version of Figure 7-4. When we analyzed Figure 7-4, we spent a whole paragraph revealing the transistors connected as NOR gates, and all of the other connections.

Figure 7-5, part (b), hides the transistors inside the NOR gates, so we can better view the connections. The diagram in part (b) helps us see that, strangely, the output of each NOR gate feeds back into the input of the other NOR gate. Also, note in part (b) that where the lines crisscross, they do not touch each other. One line hops over the other. When lines (wires or circuit board traces) are meant to connect to each other, we usually draw a little black circle.

Let's start with an overview of how this latch works. It is not our finished product, because this RS latch, though it stores, is not selectable. It doesn't have the E control of Figure 7-2. Refer to Figure 7-6, below, as you read the following summary.

Normally, the R and S inputs are both low. When we want to store a 1 (5 volts) in the latch, we raise the S (Set) input to 1. The logic 1 appears at Q. We pull the S back to 0 (0 volts), and the stored logic 1 stays at Q.

Again, normally, the R and S inputs are both low. When we want to store a 0 (0 volts) in the latch, we raise the R (Reset) input to 1 (5 volts). The logic 0 appears at Q. We pull the S back to 0, and the stored logic 0 stays at Q.

We have names for the states of the latch. The set state means the Q output is storing a 1. We can use "set" as a verb: the latch has been set to 1. The S label stands for "set."

The reset state means the Q output is storing a 0. We can use "reset" as a verb: the latch has been reset to 0. The R label stands for "reset."

The hold state occurs when both R and S inputs are 0. The latch is neither being commanded to Set, nor to Reset. So, it is holding--or storing--the last state it was commanded to be in, when there was a 1 on either R or S.

The latch should never have a 1 (5 volts) at both the R and S inputs, at the same time. Under those conditions, the latch is being told to set and reset at the same time. Q is being told to be at 0 volts and 5 volts at the same time, to be simultaneously true and false. This is known as the illegal state, and produces unpredictable results.

We can construct a brief table to summarize the behavior of our RS latch. See the picture below:

RS Latch

S	R	Q	$\overline{Q}$
0	0	Hold	Hold
0	1	0	1
1	0	1	0
1	1	?	?

Figure 7-6

Use Figure 7-6 to help you remember the content of the paragraphs immediately preceding it. As with previous logic tables, each row represent a unique output state and the S and R inputs that cause it. The question marks in line four of Figure 7-6 reflect the unpredictable results of the illegal state. Our final design will render reaching this state impossible.

Thus far, we see how a latch works from the outside: from the point of view of external inputs and outputs, and from logic tables. But, what is the secret, what is the magic trick, the sleight of hand inside that makes it work? For that, we need to peer inside the circuitry. Let's do that by looking at the picture below.

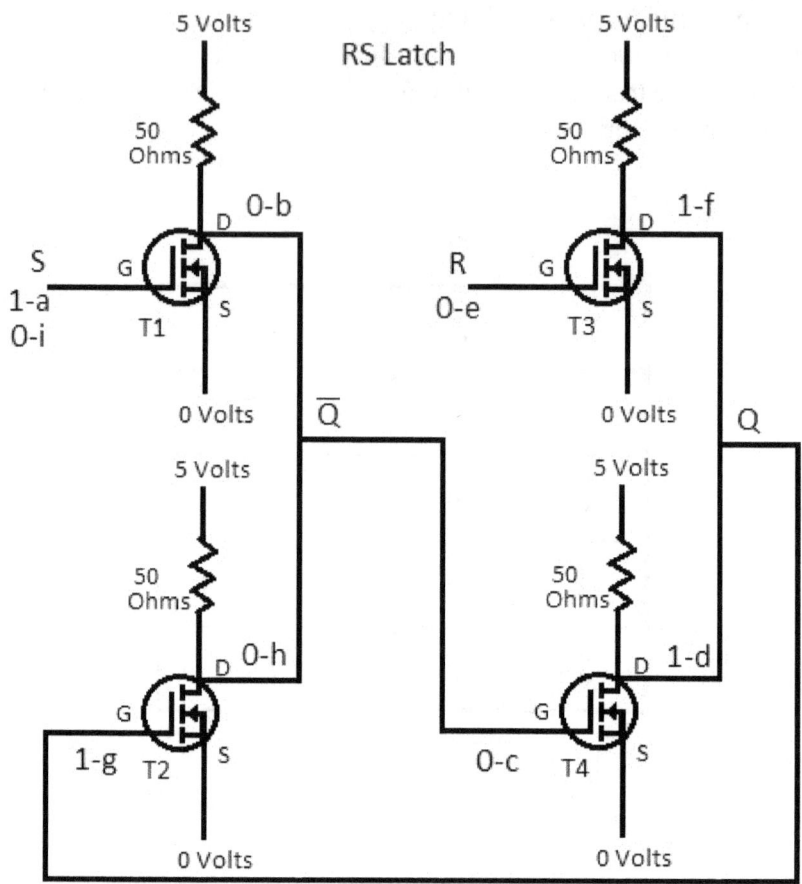

Figure 7-7

Let's demonstrate the third line of the truth table from Figure 7-6. It says that applying a 1 to the S input stores a 1 at the Q output. Hopefully, by now, you are comfortable mentally replacing a 1 with 5 volts and a 0 with zero volts as you look at Figure 7-7. I have labelled the sequence of events as "a" through "i" in the above figure. The label 1-a means that, at that location, the first step--a--of the sequence happens; and the logic level at that location is 1 (5 volts).

At 1-a, 5 volts is applied to the Set input. T1 turns on, like a closed switch, bringing T1's output to 0 volts (see 0-b). If you recall our

earlier looks at the parallel configuration of T1 and T2, you know that
it doesn't matter what voltages are around T2. "The low voltage always
wins," at the drain junction of T1 and T2. (Review Figures 4-13 and 6-
17: the off valve/transistor can't build up pressure, because current
bypasses it and heads to the on valve/transistor.) So, the NOT Q output
is at 0 volts, which looks good for the truth table row we are trying to
prove.

At 0-c, 0 volts is applied to the gate input of T4. T4 turns off, like a
closed switch, potentially bringing its drain output to 5 volts (see 1-d).
I say, potentially, because if T3 is on, its 0 volts output would over-
ride T4's 5 volts output ("the low voltage always wins" at the T3-T4
junction). However, T3 is not on. The R input to T3 is 0 volts (0-e),
keeping T3 off (1-f). Remember, it would be illegal for R to be at 5
volts when S is at 5 volts. So, the Q output is at 5 volts, which also
looks good for the truth table row we are trying to prove.

But, we still have three concerns. One, the S input of 5 volts will
go away, becoming 0 volts; how will this affect Q and NOT Q? Two,
how is the 5 volts stored at Q: by what mechanism does it "stick" there?
Three, where is the magic trick that makes this storage happen?

All three of these questions are variations of the same question,
and its answer is at step 1-g. We left our circuit at 1-d and 1f, with 5
volts there, since T3 and T4 are both off. That 5 volts is applied at 1-g,
to the gate input of T2. Here's where the "trick" happens. The 1-a step
of applying 5 volts to the gate of T1 has just been replicated at the gate
of T2 (1-g). It happens instantly, even though it took me many words.
T2 turns on (0-h), just like T1 is on. They both pull their drains to the 0
volts state. However, only one of them is required to maintain that low
voltage state. Remember, "The low voltage wins," at the outputs of
these kinds of circuits? So, the input at S into the gate of T1 can now
safely return to 0 volts (0-i), because T2 is now doing T1's job of
holding Q at 5 volts and NOT Q at 0 volts. At step 0-i, we have arrived
at the hold state, line one of the truth table, since both R and S inputs
are at 0 volts. The latch is "holding" the result of the Set state.

We could do a similar analysis of line two of the RS latch truth
table, reworking Figure 7-7 for the Reset state, by placing 5 volts at
the R input, and 0 volts at the S input. But, since the right side of the
circuit is a mirror image of the left side, the analysis would be identical,
but with opposite results: R would make Q equal 0 volts, and NOT Q
equal 5 volts. In fact, if we went to Figure 7-7 and relabeled R as S, S
as R, Q as NOT Q, and NOT Q as Q, we could use the same diagram

and the same analysis to describe line two of the truth table.

Instead, let's use Figure 7-5, part (b), a higher-level abstraction, to analyze line two of the RS latch truth table. Look at the picture below:

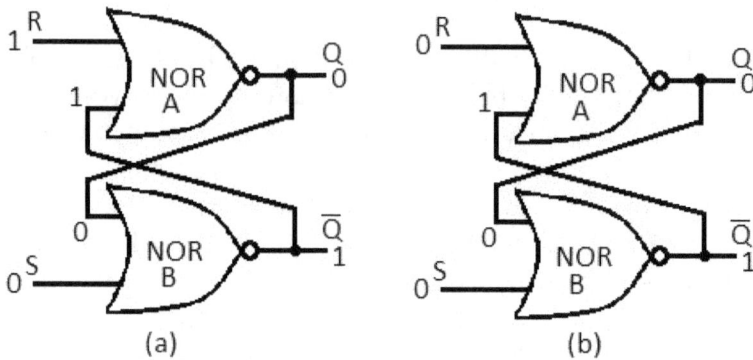

(a)   (b)

FIGURE 7-8

Figure 7-8, part (a), displays the act of resetting the RS latch. (Remember, logic 1 means 5 volts for our design, and logic 0 means 0 volts.) 5 volts is applied to the R input of the latch, and enters NOR A. Refer back to the logic table for the NOR gate, found in Figure 4-20, part (b). The truth table tells us that if one input of a NOR gate has 5 volts on it, the output is automatically 0 volts. It doesn't matter what the other input is. So, NOR A's output, which is the Q output of the latch, is 0 volts. That is the expected result for the Reset operation: Q is reset to 0 volts, by definition.

Notice that that same 0 volts feeds back to one input of NOR B. The other input of NOR B is also 0 volts. The truth table tells us that if all inputs of an NOR gate are 0 volts, the output must be 5 volts. So, 5 volts comes out of NOR B. That is the expected result for the Reset operation: NOT Q is 5 volts, the opposite of Q's 0 volts, by definition. So far, so good.

But, the R input must go back to 0 volts, in case the latch needs to be Set in the future. It can't be Set by applying 5 volts at S while 5 volts are still at R: that would be the illegal state, with unpredictable results.

Now, we can observe the "trick" that maintains and stores the result of the Reset state. Still looking at part (a), notice that the 5 volts that we analyzed as coming out of the output of NOR B feeds back to the other input of NOR A. That means that NOR A now has both inputs at 5 volts. The truth table tells us that if both inputs of NOR A are at 5 volts, 0 volts comes out of NOR A. But, NOR A doesn't need both inputs to be at 5 volts, it only needs one input to be at 5 volts to produce a 0 at its output.

It's safe, then for the Reset input, R, of the latch to return to 0 volts. The R input of NOR A no longer needs to be 5 volts, because the other input of NOR A is now at 5 volts. That's the "trick" that maintains Q at 0 volts and NOT Q at 5 volts. You can observe this in Figure 7-8, part (b). The R input is now back to 0 volts. The S input always was at 0 volts. We have entered the hold state, and the latch is holding or storing the result of the Reset state. Though it took many words to describe the replication of the 5 volts from the R input of NOR A into NOR A's other input, it happens very quickly. The R input only needs a brief pulse of 5 volts to store the 0 volts at the Q output.

Our latch design works well, but it is unfinished. It needs circuitry built into it that allows it to be selectable, as we witnessed in Figure 7-2, part (a). Our latch may be one of many latches, all with their Set bits connected together, and all with their Reset bits connected together. They all may be connected to one source bit of information. The source must be able to select just one latch to send information to. The correct latch must be able to receive it, and the other latches must be able to ignore the information not meant for them. Let's look first at the picture below:

D Latch

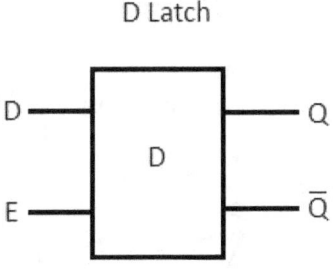

E	D	Q	$\overline{Q}$
0	X	Hold	Hold
1	0	0	1
1	1	1	0

(a)   (b)

FIGURE 7-9

Figure 7-9, part (a), is our high level logic symbol for our finished latch design. Notice, it has a new name. It is now a D latch, instead of an RS latch. The D stands for delay.

The D latch has two inputs, D and E, and 2 outputs, Q and NOT Q. NOT Q is always in the opposite logic state of Q, so I'll ignore it, to save time. The truth table in part (b) explains how the D latch works. Again, we look at separate behaviors of a truth table one row at a time. Again, a logic level of 1 means, for our design, 5 volts; a 0 logic level means 0 volts.

We introduced the letter E in Figure 7-2. E stands for enable. E is the voltage input that enables the D input. When the E input is 5 volts, as seen in rows two and three of the truth table, the D input voltage appears at the Q output. So, 0 volts on D appears as 0 volts on Q (truth table row two). And, 5 volts on D appears as 5 volts on Q (truth table row three). I repeat, these transfers only occur when the E input is at 5 volts.

What happens, then, when E is 0 volts? The "x" in row one of the truth table means, "don't care." It means, we don't care what input is on D, because the D input is ignored, it is blocked. The door is closed. The D input voltage does not cross over to the Q output. The Q output holds whatever voltage was last stored there when E was 5 volts.

So, whatever is hidden inside the box that the E line connects to in Figure 7-9, part (a), holds the secret to our latch's ability to be correctly selected and stored to by the source of information. Let's open up that box. Its mysteries are revealed in the picture below:

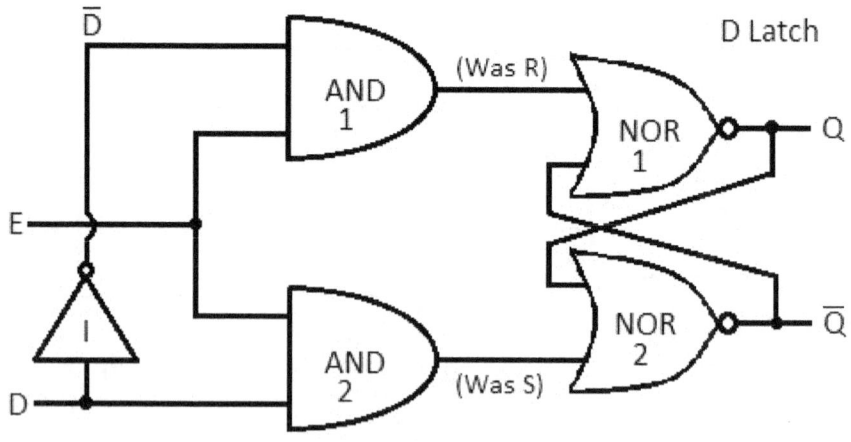

FIGURE 7-10

Figure 7-10 is our finished latch. Notice first that our old RS latch is on the right side: NOR 1 and NOR 2, connected the same way. So, we know our new, D latch can store. On the left is new circuitry, which allows our D latch to either be selected or ignored. Notice how I have changed the labelling of R to "was R," and S to "was S". Their behaviors haven't changed, but the name of the latch has. This is a D latch, so we really should not be using RS labels. You will notice we now have a D external input on the far left side, which replaces the old R and S external inputs. Also, we have our new control input, called E. You know its purpose from the above discussion.

We already know how the right side of Figure 7-10 works. 5 volts at "Was S" sets Q to 5 volts; 5 volts at "Was R" resets Q to 0 volts. 0 volts at both "Was S" and "Was R" makes Q hold or maintain the previously stored voltage bit.

Let's look at that inverter, labeled I, on the left side of Figure 7-10. The inverter allows us to compress two inputs, R and S, down to one input, D. Look back at Figure 7-6, the truth table for the RS latch. Notice that in rows two and three, the R and S inputs are in opposite states from each other. In other words, in row two, S is 0 and R is 1: opposites. In row three, S is 1 and R is 0: opposites. We don't need rows one and four. Row four is illegal, and a source of potential trouble. Our E input, a separate circuit, will perform the hold state, so we don't need row one.

An inverter always has its input and output opposite each other. So, we wire it as shown in Figure 7-10 to replace two inputs, R and S, with one input, D, which has an inverter output of NOT D. The D input takes on the role of the old S input. The NOT D output of the inverter takes on the role of the old R input. That leaves us to explain how the E circuit takes care of the hold state's functionality.

The task of the E circuit is to block input data D from getting to "Was S" and "Was R". When E is at 0 volts, the E circuit must keep "Was S" and "Was R" at 0 volts, regardless of the voltage at D. When "Was S" and "Was R" are both at 0 volts, NOR 1 and NOR 2 will be in the hold state.

The E circuit accomplishes this task via the AND 1 and AND 2 gates. One way of looking at an AND gate is as a "pass/block" circuit. Look at the picture, below:

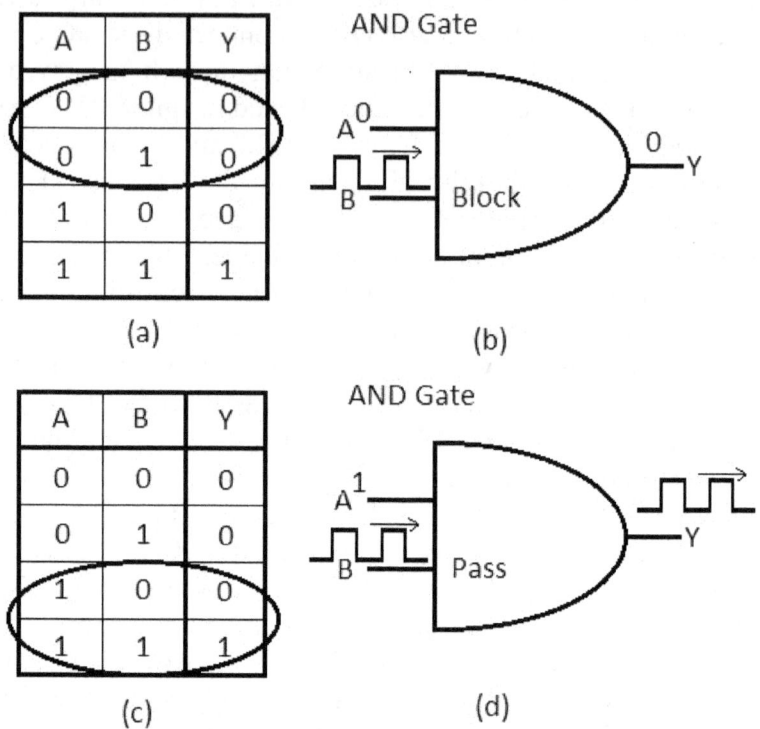

(a)

(b)

(c)

(d)

FIGURE 7-11

Look at the AND gate truth table in Figure 7-11, part (a). I have circled the first two rows. In both these rows, the A input is 0 volts. Notice that, no matter what the value is at the B input, the Y output is 0 volts. You can watch this behavior playing out in part (b). The A input is 0 volts; the B input is changing from 0 volts to 5 volts, repeatedly. Nevertheless, the Y output stays stuck at 0 volts. The B input is blocked from reaching Y, when A is 0 volts.

Now look at the AND gate truth table in Figure 7-11, part (c). I have circled the last two rows. In both these rows, the A input is 5 volts. Notice that whatever voltage is at the B input appears at the Y output. You can watch this behavior playing out in part (d). The A input is 5 volts; the B input is changing from 0 volts to 5 volts, repeatedly. The Y output exactly matches the B input. When B is 0 volts, Y is 0 volts. When B is 5 volts, Y is 5 volts. The B input passes through to the Y output, when A is 5 volts.

This "pass/block" functionality is exactly what we want in our latch design. Look once more at Figure 7-11. The way the A input controls the information flow is exactly what we want our E input to do.

Now, look back at Figure 7-10, our actual D latch circuit. Think of the E input as the controller of information flow to "Was R" and "Was S." When E is 5 volts, it puts AND 1 and AND 2 in the "pass through" mode. Then, if D is 5 volts, AND 2 lets it pass, and "Was S" becomes 5 volts. Also, the inverter makes NOT D 0 volts, so AND 1 lets it pass, and "Was R" becomes 0 volts. We are now at the right side of Figure 7-10, which is our old RS latch, with voltages that match row three of its truth table (Figure 7-6). Of course, this means Q becomes 5 volts.

Again, let's look at the scenario where E is 5 volts, putting AND 1 and AND 2 in the "pass through" mode. This time, though, assume D is 0 volts. AND 2 lets it pass, so "Was S" also becomes 0 volts. Also, the inverter makes NOT D 5 volts, and AND 1 lets it pass, so "Was R" becomes 5 volts. We are now at the right side of Figure 7-10, which is our old RS latch, with voltages that match row two of its truth table (Figure 7-6). Of course, this means Q becomes 0 volts.

Lastly, let's look at the scenario where E is 0 volts. Now, E puts both AND 1 and AND 2 in the blocked state. Both their outputs become stuck at 0 volts. We are now at the right side of Figure 7-10, which is our old RS latch, with voltages that match row one of its truth table (Figure 7-6). Of course, this is the hold state, which means the Q output merely maintains what previous data it held, regardless of what voltage is at D or NOT D.

Now that you know how the D latch works internally, let's look at how the microprocessor uses latches. We introduced these concepts in Figure 7-2, part (a), to explain the need for selectable storage. Then, we used terms like source device and destination device. Now, let's review this process more realistically, using D, Q, and E labels from our latch tutorial, and from the point of view of a microprocessor communicating with its latches.

Assume there are two D latches: latch A and Latch B. The picture below may help:

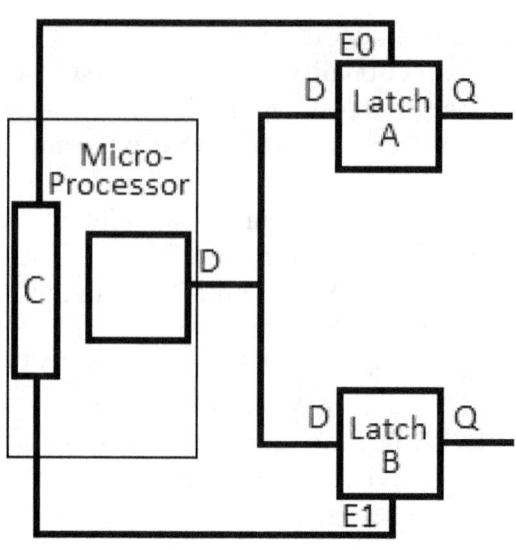

FIGURE 7-12

If the microprocessor wants to send a 1 to latch A, it places 5 volts on the D line (wire/trace). Then, the microprocessor raises latch A's E0 line to 5 volts. Then, the microprocessor lowers the E0 line to 0 volts. This E0 line is uniquely for latch A; it is for no other device. Latch A now stores the logic 1, delivering 5 volts to its own Q output.

Next, assume the microprocessor wants to send a 0 to latch B. It places 0 volts on the D line (wire/trace). Then, the microprocessor raises latch B's E1 line to 5 volts. Then, the microprocessor lowers the E1 line to 0 volts. This E1 line is uniquely for latch B; it's for no other device. Latch B now stores the logic 0, delivering 0 volts to its own Q output.

After these two step are complete, latch A is storing a logic 1, and latch B is storing a logic 0. Note that both latch A and latch B share the same, one, wired connection called D, from the microprocessor. However, latch A and latch B do not share the E line. Each latch gets chosen by its own, unique E line.

As we have described earlier, as in Figure 7-3, part (a), and its associated text, in a microprocessor system, we usually want to store many bits at the same time, rather than one bit at a time. Instead of sending to latch A, then later to latch B, we usually send to latch group

A, then to latch group B. It's not unusual for a latch group consist of 64 bits.

Let's design a 4-bit D latch. An 8-bit, 16-bit, 32-bit, or 64-latch would look similar. See the picture below:

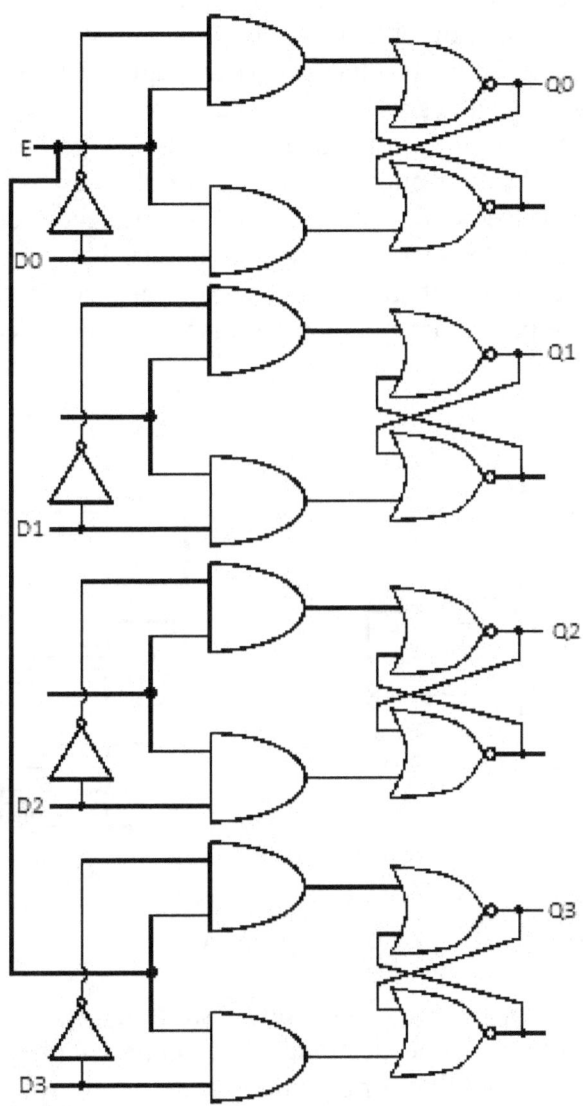

FIGURE 7-13

This 4-bit D latch is simple to design and understand. You should recognize our 1-bit D latch repeated four times. All four of the E lines are tied together, and connected to the one E input line. The four D lines, however, are kept separate. Thus, each D latch in the group can store its own bit of information, received from its own D input. But, all four can be loaded at the same time, from one E line. Notice, too, the four available D latch outputs, Q0 through Q3. With this design, if Q0 is the only output that must change, Q1 through Q3 must be re-loaded with their previous values when Q0 is loaded with its new value.

Another consideration for our designs of both gates and latches is the effect of connecting outputs together. Look back, now, at Figures 7-1 through 7-3. Remember, for reasons of efficiency, detailed in the paragraphs around those figures, we found we had to accommodate one source of output voltage connecting to many destinations of input voltage. The opposite arrangement is also common. See the picture below:

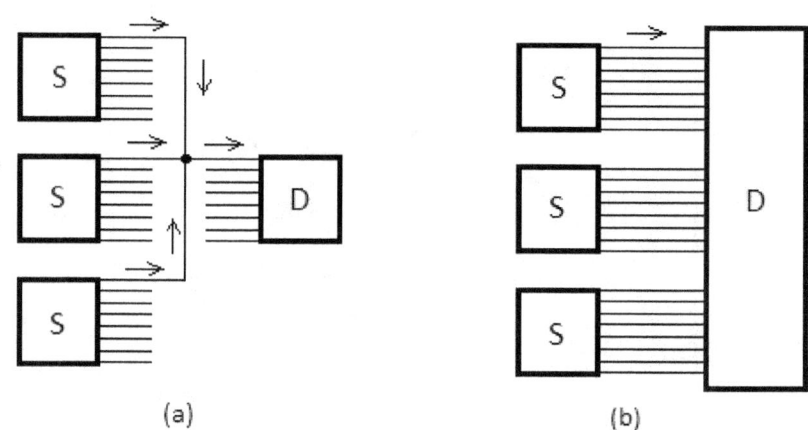

(a)   (b)

FIGURE 7-14

This new, opposite, arrangement is: many sources of output voltage connecting to one destination of input voltage. Figure 7-14, part (a), demonstrates that arrangement. Part (b) shows what problem part (a) overcomes. If we could not select and store with latches, the destination would need a line for every source line, 24 lines in this example. If each source had 64 bits, and there were ten sources, the destination would need 640 lines (wires or traces).

Figure 7-14, part (a), shows three 8-bit latch sources (S) connecting to one 8-bit latch destination (D). Instead of 24 lines, the destination (D) in part (a) only requires eight lines. The destination (D) needs to have a copy of the voltages contained in only one of the three sources (S).

Not shown in Figure 7-14, part (a), is the control circuitry necessary to make sure that only one source (S) at a time is selected to deliver a copy of its 8 bits to the destination (D), while the other two sources are turned off. I have also not drawn all of the line connections, to prevent chaos and confusion. But, I did draw one of the eight connections. Follow the arrows to see the direction of voltage transfer.

But, how do we turn on or off output voltages in the sources? Unfortunately, with the circuit we have been using thus far--our basic transistor/resistor inverter--*it can't be done!* 0 volts stored at any "on" output transistor would defeat the 5 volts stored at any "off" transistor. When connected together, output voltages contaminate, and in some designs destroy, each other.

We need a re-design at the exit of any gate or latch with outputs tied together. The re-design will enable the gate or latch to effectively disappear, from the point of view of other gates or latches. The one latch or gate that does not "disappear" can then send voltages to the destination, without contamination or destruction.

We need a second type of transistor for this new design. Look at the picture below:

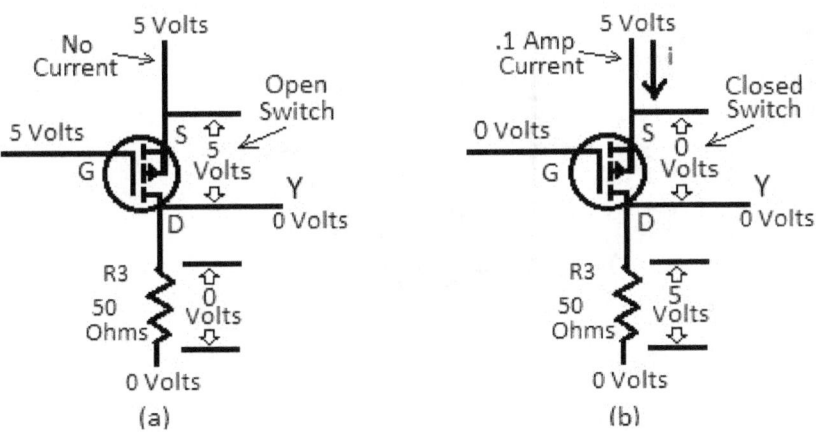

FIGURE 7-15

The transistor in Figure 7-15 is called a P-channel enhancement-mode MOSFET. Compare this with the one we have been using, the N-channel enhancement-mode MOSFET, as seen in Figure 6-1. You will find that this new one is like a vertical mirror image of the old one. Notice that I have drawn this new one above the resistor; I have placed the source (S) at the top and the drain at the bottom; the arrow inside the circle points outwards.

The most important things to notice are how the voltages and currents are reversed, compared to the old N-channel transistor. Look at Figure 7-15, part (b). The conventional current flows from 5 volts to ground, as before; but now, it enters the source and exits from the drain. Also, to turn on the source-to-drain current, the gate must now be made more negative (0 volts) than the source (5 volts). The source-to-drain then acts as a closed switch, with 0 volts across it.

Now, look at Figure 7-15, part (a). To turn off source-to-drain current, the gate must now be made the same (5 volts) as the source (5 volts). The source-to-drain then acts as an open switch, with 5 volts across it. All these behaviors are the reverse of what we encountered with the N-channel transistor.

Let's now look at a new inverter design. Look at the picture below:

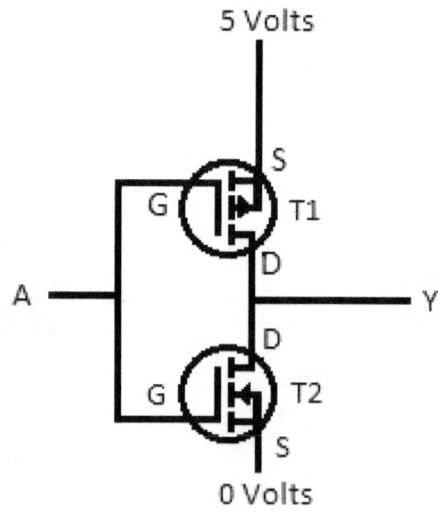

FIGURE 7-16

The first thing to notice in our new inverter design is that the old resistor has been replaced by a second transistor, T1. T1 is a P-channel enhancement-mode MOSFET. T2 remains unchanged; it is an N-channel enhancement-mode MOSFET. Notice also that their gates are wired together, then wired to the input voltage, A. Finally, notice that their drains are wired together, then wired as the output.

We could have built all of our gates and latches with two transistors, an N-channel and a P-channel enhancement-mode MOSFET, wired in this way. I chose not to, for two reasons. One, I wanted to teach through a comparison of transistors with water faucets, and water faucets do not work this way. Second, this circuit is harder to understand, involving two different types of transistors. Understanding it depends more on electronic theory than hydraulics. You should be more comfortable now with the underlying electronic theory. You'll see shortly why we now need this circuit.

Look at the picture below, so we can explore how this new inverter design works:

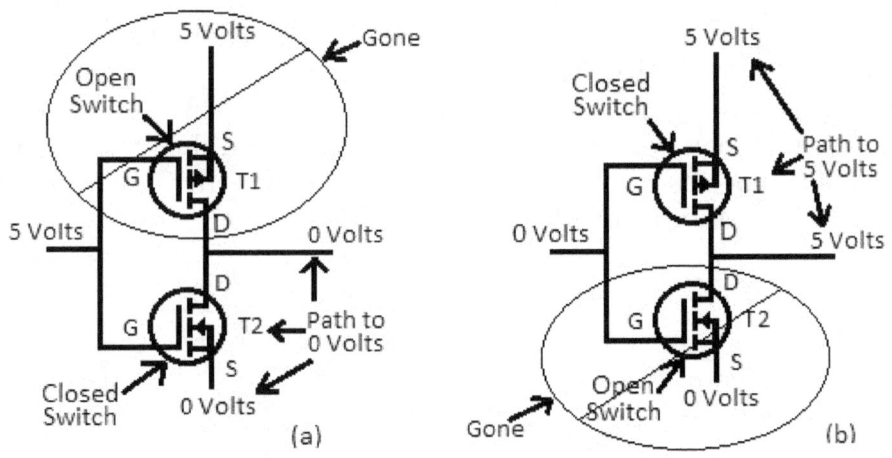

FIGURE 7-17

I have drawn the same circuit twice, so we can study its behavior with 5 volts input, then with 0 volts input. The first thing to notice about Figure 7-17 is that this circuit truly is an inverter. In part (a), 5 volts comes in, and 0 volts comes out. In part (b), 0 volts comes in, and 5 volts comes out. Let's see how this happens.

In part (a), 5 volts is applied to both gates. T1--the P-channel transistor--turns off, since its gate voltage is the same as its source voltage. It acts like an open switch. Recall that an open switch is nothing but air. That part of the circuit effectively disappears, as I have indicated in part (a) with the crossed-out circle.

Still in part (a), the 5 volts applied to the gate of T2, the N-channel transistor, turns it on. Drain to source, it acts like a closed switch, like a wired path to neutral (0 volts). That 0 volts appears at the output. The 5 volts at the source of T1 is totally blocked.

Turning now to part (b), 0 volts is applied to both gates. T2, the N-channel transistor, turns off, since its gate voltage is the same as its source voltage. It acts like an open switch. Recall that an open switch is nothing but air. That part of the circuit effectively disappears, as I have indicated in part (b) with the circle.

Still in part (b), the 0 volts applied to the gate of T1, the P-channel transistor, turns it on. Drain to source, it acts like a closed switch, like a wired path to 5 volts. That 5 volts appears at the output. The 0 volts at the source of T2 is totally blocked.

There is, however, a third thing we can do with this inverter that we could not do with our resistor/ transistor inverter. To see this feature, look at the picture below:

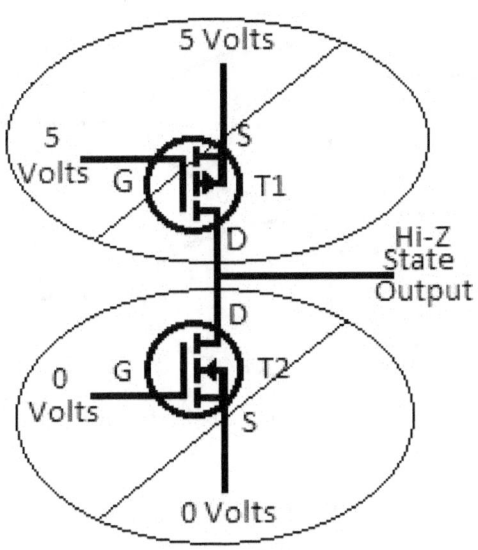

FIGURE 7-18

In Figure 7-18, I have removed the connection between the two gates. I now can control the two transistors separately. So, I apply 5 volts to the gate of T1, turning off the P-channel FET. From the point of view of the output, T1 is gone. Next, I apply 0 volts to the gate of T2, turning off the N-channel FET. From the point of view of the output, T2 is gone. In fact, from the point of view of the output, the whole inverter is gone! This disappearing act is sometimes referred to as the hi-Z state, meaning high impedance state, although it reminds one of the circuit going to sleep, as in "catching some Z's."

This is how we can have many digital voltage outputs connected together without the voltages affecting each other. We simply make unused outputs disappear by putting them in the hi-Z state. Figure 7-18, above, shows us a way to make outputs disappear.

Our work is not yet done, though. We need some additional controls so that, when this circuit is active, it behaves like Figure 7-17; and when this circuit is disabled, it behaves like Figure 7-18. See the picture below, for the finished design.

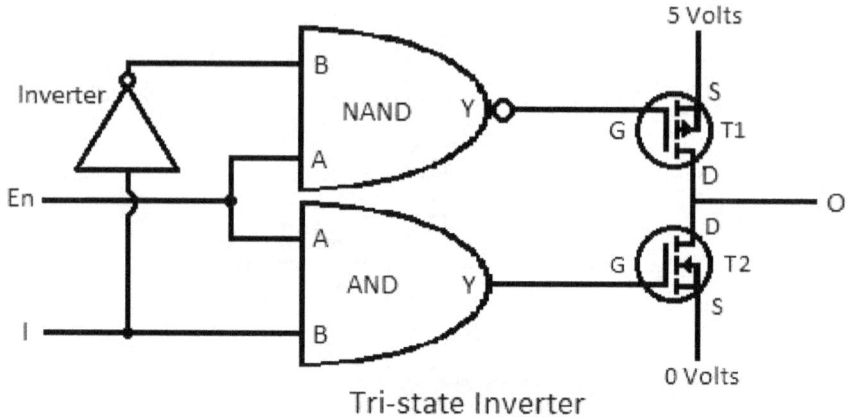

Tri-state Inverter

FIGURE 7-19

The circuit in Figure 7-19 is often called a tri-state inverter. It's called tri-state because its output can be in three different states: 0 volts, 5 volts, and the hi-Z state described above.

When the En input is at 5 volts, the circuit in Figure 7-19 behaves like a normal inverter. En stands for Enabled, and it means that the output voltage, at O, is able to reach a destination circuit. Either T1 or T2 is turned on. On the other hand, when the En input is 0 volts, the output voltage, at O, is unable to reach a destination circuit. Both T1 and T2 are turned off.

Let's start with the latter, first. When En is low (0 volts), the A inputs to both the AND and the NAND gates are low. The truth table for an AND gate, as seen in Figure 6-22, shows that when the A input is low, the Y output is low. You don't even have to bother looking at the B input. This 0 volts turns off T2. Also, the truth table for a NAND gate, as seen in Figure 6-26, shows that when the A input is low, the Y output is high (5 volts). You don't even have to bother looking at the B input. This 5 volts turns off T1. Both T1 and T2 are off when En is low, so this circuit behaves like Figure 7-18: the output voltages no longer affect circuits they are attached to.

Now, let's see what happens when En is high (5 volts). This is a little trickier. What we would like is that, when En is high, the I input voltage level gets applied to the gates of both T1 and T2. This way, it would be acting just like Figure 7-17: like an inverter.

Here's what actually happens. When En is high (5 volts), a high logic level is applied to the A inputs of both the AND and the NAND gates. From our earlier discussion about Figure 7-11, we remember that for an AND gate with the A input high, the B input (from I) passes through and becomes the Y output. This is what we want. The I input becomes T2's gate input.

A careful look at the NAND gate logic table in Figure 6-26 reveals that for a NAND gate with the A input high, the B input passes through, *flips to the opposite logic level*, then becomes the Y output. This is not what we want. *We don't want this flip*. If we could flip this flip, then the I input's voltage would be the same as T1's gate voltage.

That's the function of the inverter. The inverter flips the I input to the opposite logic state. But, the NAND (with the A input high) flips the I logic state back to its original value as it passes from its B input to its Y output. Thus, with En high, the I input value does, in fact reach both T1's and T2's gates. With En high, our Figure 7-19 circuit behaves like an inverter, and the output voltage at O affects the destination circuit. With En low, our Figure 7-19 circuit's output at O is completely disconnected from any destination circuit.

If you don't want the circuit in Figure 7-19 to invert the input logic as it passes from I to O, the solution is simple. Apply the I input to B of the NAND. Then, attach the I input to the input of the inverter, point the inverter downwards, and attach the output of the inverter to B of the AND gate. See what I mean by looking at the picture, below, then comparing it to Figure 7-19. If the tri-state circuit does not invert, it is called a tri-state *buffer*.

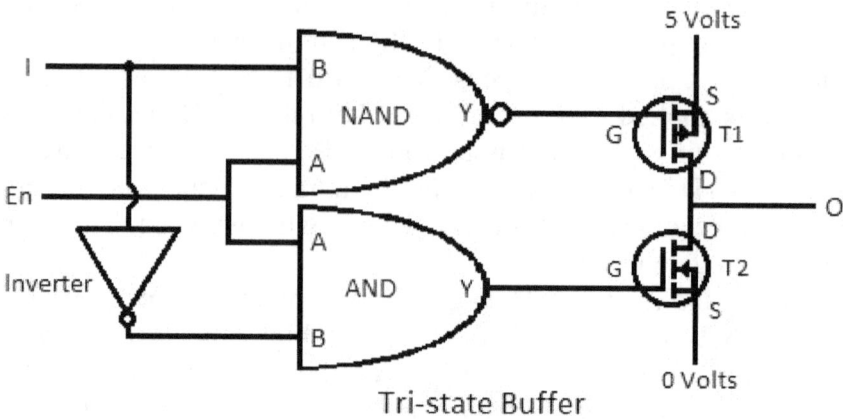

Tri-state Buffer

FIGURE 7-20

Our next topic is another stunning one. It is certainly amazing that we are able to move voltages from one part of a digital system to another. Or, think of what is going on at the machine level when we add two numbers together. A group of binary voltages is moved to the A input of an adding circuit; a second group of binary voltages is moved to the B input of an adding circuit; the circuit produces the sum of A and B as a group of binary voltages at its Y output. All of this is accomplished with the same two electrical components arranged as our inverter; then those inverters are combined in patterns we call gates and latches; then those gates and latches are arranged in patterns we call digital adders.

It is certainly amazing that we can move a copy of voltages from source A to destination B. But, *we also move them back, along the same wires, from B to A*. We can do this with one latch, but usually we do this with a group of latches. Think of this. We can move a voltage pattern, in logic form as 10101010 from 8-bit latch A to 8-bit latch B, along eight wires. Some other circuitry can modify that value (maybe subtracting from it) in B. The resultant, new voltage pattern can be moved back to A, in the reverse direction, along the same 8 wires!

Now, look back at Figure 7-3. It shows one source latch group able to pick out one of many destination latch groups, and "write" to it. We define the write operation as the copying of a bit pattern from the controlling latch group to a destination latch group. Next, look back at Figure 7-14. It shows one destination latch group able to pick out one of many source latch groups, and "read" from it. We define the read operation as the replicating or copying of a bit pattern from a source latch group to the controlling latch group.

The ability to copy voltages in both directions on the same wires gives us the power to *combine* the capabilities demonstrated in figures 7-3 and 7-14. In Figure 7-3, the source latch group was the controller. In Figure 7-14, the destination latch group was the controller. In a microprocessor-based system, the microprocessor is the controller. It is both the source and destination of information flow. It "reads" voltages into itself, and it "writes" information elsewhere. Look at the picture below, in part (a), showing a complex system in which voltages can move both ways.

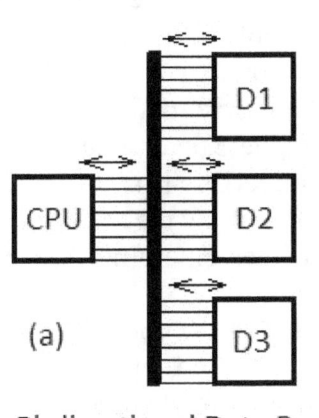

Bi-directional Data Bus

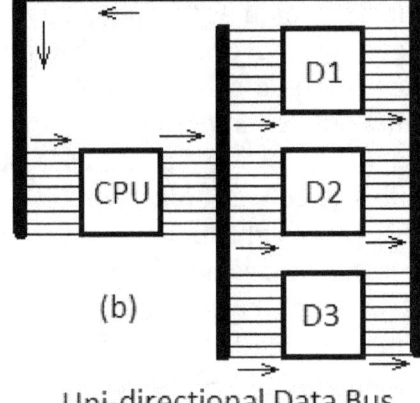

Uni-directional Data Bus

FIGURE 7-21

Ignore part (b), for now. Looking at Figure 7-21, part (a), we need to define some terms and concepts. First, D1, D2, and D3 are 8-bit latch groups. The component labelled the "CPU" is the central processing unit: the "brains" of the microprocessor. However, the internal part of the CPU attached to the 8 wires going to D1 through

D3 is just another 8-bit latch. We will refer to it as the CPU latch.

Next, a "bus" is a collection of wires with a common purpose. Voltages move simultaneously, or "in parallel," across these wires. We can call the eight wires in Figure 7-21, part (a), a "data" bus, since the group of eight voltages on it is information or "data."

Next, "bi-directional" means voltages can move in either direction along the bus: from the CPU latch to storage in one of the three D's; or from one of the three D's to storage in the CPU latch. Each of the D's is an 8-bit latch, or latch group.

Don't think of the D latch groups as only being destination latches anymore. The D now stands for the fact that they are built from D-style latches. In earlier figures, a latch group labelled S was always a source latch group, an output of voltages; while a latch group labeled D was always a destination latch group, a destination for voltages. In a bi-directional bus system, any latch group is sometimes a source, and at other times a destination.

In Figure 7-21, part (a), we also need to describe some new drawing techniques. The arrow with a tail on both ends signifies that voltages can move in both directions along the bus. The thick, black, vertical line is a shortcut way of drawing the 8-wire data bus. There are truly only eight unique wires in our system, here, even though each wire splits three ways as it heads out to the three D's. It is difficult to draw this three-way split eight times, so we often draw the thick line to stand for the eight lines, then draw the eight lines emerging from the thick line. There is no "thick wire."

If you look back at Figure 7-3, or at Figure 7-13, you will notice that, in a standard latch, for every input line, we need an output line. Notice in Figure 7-21, part (a), I have not drawn eight lines on the right side of each D latch group. I don't need to draw them, because they don't exist: each input and output shares the same line. Without the bi-directional bus, I would need twice as many wires: I would need the additional eight wires coming out of the right sides of each of the D1 through D3 latch groups, then merging to become eight unique wires, then entering into new input connections on the left side of the CPU. Figure 7-21, part (b), depicts the messy complexity I have just described.

If we had not acquired the capabilities of the one-to-many in Figure 7-3; the capabilities of many-to-one in Figure 7-14; and the capability of using one bus instead of two separate ones for moving data into and out of the CPU, our system in Figure 7-21 would require

48 unique wires, instead of 8. A "real world" example with 10 peripheral latches and a 64-bit data bus requires 64 unique wires, if implemented *with* all our design tricks. *Without* all our design tricks, the "real world" example would require 1280 wires.

Let's now look at the circuit that allows us to have a bi-directional bus. See the picture below:

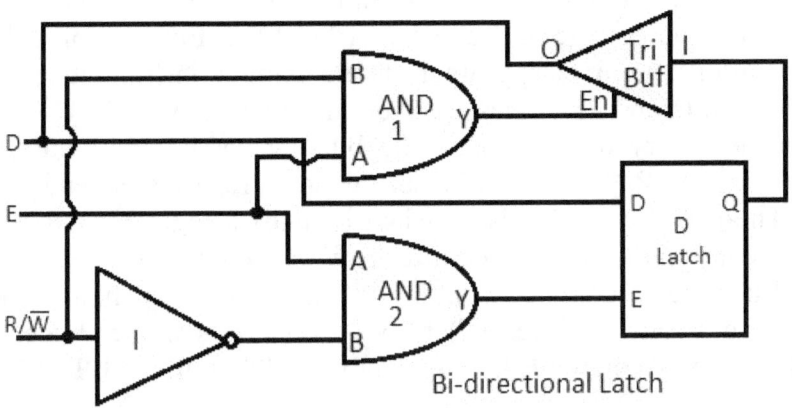

FIGURE 7-22

This circuit displays how to turn a normal, 1-bit latch into a bi-directional latch. You would combine eight of these together to make an 8-bit latch, like D1 in Figure 7-21, part (a). All eight would share one E and one R/W line.

Let's start with the box labeled "D-Latch" in the lower right-hand corner. This is the same select-and-store circuit we studied in Figure 7-10. The rest of the circuitry in the picture allows the D input and Q output to share the same wire, or turns off access altogether.

The D input and Q output share the same wire by taking turns on the same wire. When the D voltage comes into this circuit, the latch's Q voltage output must be blocked. When the Q voltage comes out of this circuit, the latch's D input must be blocked.

How do we stop the Q output from reaching the D line, the data bus line? We already know how. We already designed a tri-state buffer so that many outputs can be connected together, turned off, and only one turned on at a time. In the design of Figure 7-22, we need to turn off the Q output when its voltage would contaminate the voltage coming into the D latch input. Thus, you will see the tri-state buffer drawn in the upper right-hand corner of Figure 7-22. This is your first look at the logic symbol for a tri-state buffer. Inside, it contains the circuit you saw in Figure 7-20. When the En line on the tristate buffer is 5 volts, the voltage on I appears at O. When En is 0 volts, the tri-state buffer's output disappears, from the D input line's point of view.

The D, E, and R/W lines are under the control of the CPU. In short, D is the binary data bit, E is the bit that enables this latch to do anything, and R/W selects whether this latch receives or sends a logic bit. The CPU reads this latch when it sends a logic 1 on R/W. The CPU writes to this latch when it sends a logic 0 on R/W.

When the CPU's latch want to write a bit to this latch, it makes the R/W line 0 volts. It places either 0 volts or 5 volts on the D line. Then, it places 5 volts on the E line. It can then safely pull the E line back to 0 volts.

When the CPU's latch want to read a bit from this latch, it makes the R/W line 5 volts. Then, it places 5 volts on the E line. It grabs the latch's data off the D line. It can then safely pull the E line back to 0 volts. Thus, R/W controls the direction of information transfer on the D wire.

Let's look at how the circuitry in 7-22 accomplishes all of this. Notice, first, that both the latch and the tri-state buffer have enable control inputs: En for the tri-state buffer, and E for the D latch. Also, both En and E inputs are high-true, meaning that a high logic level turns each one on. Next, notice that En for the tri-state buffer is controlled by AND 1, and E for the D latch is controlled by AND 2. This leads to the conclusion that a logic 1 out of AND 1 turns on the tri-state buffer, for the reading of Q by the CPU. Similarly, a logic 1 out of AND 2 turns on the D latch, for writing (storing) into the D latch by the CPU.

So, if this bit latch is not the one selected by the CPU controller latch, both AND 1 and AND 2 outputs must be at a low logic level. Neither the tri-state buffer for output nor the D latch for input will be selected. Recall, from the logic table for the AND gate, as seen in Figure 6-22, that if one input to an AND gate is low, its output is low. You don't even need to look at the other output. Well, the E input on the far left gets applied to the A input on both AND 1 and AND 2. So if the external E input from the CPU controller is low, both AND gates outputs will be low, and both the tri-state buffer and the D latch will be off: that is, blocked out of the outside world on the D wire.

You may also remember (again from its logic table) that, if one input to a two-input AND gate is high, then the other input passes through to the output with the same logic level. The E input on the far left gets applied to the A input on both AND 1 and AND 2. So, if E is high, both AND 1 and AND 2 are primed for their Y outputs going high. All either one needs is a high logic level on its B input. Let's focus next on getting a high logic level to the B inputs.

But first: we don't want to allow *both* B inputs of AND 1 and AND 2 to be high. This would cause a simultaneous read/write, input/output clash, rather than the taking of turns in sharing the D bus. Our slick way to prevent this scenario is to use the inverter, seen in the lower left-hand corner of Figure 7-22. By definition, the logic levels on the input and output of an inverter are never the same; they are always opposite. Both cannot be high. So, we connect the input side of the inverter to B of AND 1, and the output side of the inverter to B of AND 2, knowing they will never both be high at the same time. We will never read and write at the same time. When E is high, we will either read or write, but never both.

Next, we connect the R/W control signal from the CPU to the input side of the inverter (and to the B input of AND 1). Assume now that E is high. When R/W is high, the Y output of AND 1 goes high, causing the tri-state buffer to turn on, passing the D latch's Q output to the bi-directional bus, so that the CPU latch can read it. The inverter inverts the high on R/W and sends a low to B of AND 2, pulling the Y output of AND 2 low, and turning off the D latch from an unwanted write.

Still assuming that E is high, let's see what happens when R/W is low. AND 1's Y output goes low, so the tri-state buffer turns off. The Q output is blocked from reaching the D line of the tri-state bus. The inverter inverts the low on R/W and sends a high to B of AND 2, pulling the Y output of AND 2 high, and turning on the D latch for storage of the input data bit on the D line.

Let's say a few words about the CPU latch. CPU, as previously stated, stands for central processing unit. It is the "brains" of the microprocessor. It is the circuitry that controls other circuitry. It is the circuitry that executes the program commands. In some microprocessor designs, the microprocessor includes only the CPU.

Non-CPU things, like memory and input/output ports, are external to the microprocessor. In other microprocessor designs, the microprocessor includes everything: CPU, memory, input/output/, etc.

In the last few paragraphs and figures, I have referred to the CPU latch. This is just one element in the CPU. It is one latch group inside the CPU: for our design it is an 8-bit latch, no different from D1, D2, or D3 in Figure 7-21.

You may have wondered how the CPU latch works, compared to latches D1 through D3 that we have been explaining in diagram 7-21. It works the same way, with one major difference: the CPU makes the R/W line on the CPU latch the opposite logic level as on the R/W line to the external latches D1, D2 or D3. This makes sense. If the CPU is trying to *read* from 8-bit latch D1, then that means it is also trying to *write* D1's contents to its CPU 8-bit latch. If the CPU is trying to *write* to 8-bit latch D1, then that means it is also trying to *read* out the contents of its CPU latch to put it in D1. Remember, the only difference between reading and writing to a latch is the logic level of the R/W line.

CHAPTER EIGHT
A MICROPROCESSOR IN ACTION

Thus far, we have been slowly building up the pieces of the puzzle that help us understand how a microprocessor works. Now, let's take a big step forward, and look at how those pieces fit together to make a working microprocessor.

Our pattern has been to proceed from actual circuits to simpler logic diagrams that encapsulate the actual circuits inside. This has allowed us to build more complex circuits by connecting logic diagrams, so that the detail of actual circuits does not overwhelm us.

Figure 7-22, above, is a good example of this. It is an interconnection of five logic diagrams. But, each logic diagram contains inside it a physical circuit which we have previously designed and studied. If we had drawn all the physical circuitry contained in these five logic diagrams, the complexity might overwhelm us.

We are now taking a big step forward to a high-level logic diagram of the whole microprocessor. It is more of a sketchy block diagram. We can use it to get a feel for how the whole microprocessor goes about doing its work. Once we get a get a feel for the functionality at this level, we will then backtrack and explore the circuits inside these blocks in more detail. Those circuits are the final pieces of the puzzle that accomplish the events witnessed in the following demonstration.

Let's begin. If someone asks you, "What does a microprocessor do?" you can answer, "Fetch, decode, and execute." It performs this sequence of actions, in this order, over and over, millions of times per second. When the execute action is complete, the next fetch begins: fetch, decode, execute, fetch, decode, execute, fetch, decode, execute, etc. When you turn on the main power switch, the first thing the microprocessor does is: fetch.

In the *fetch* operation, the microprocessor *reads the* next *command* or instruction from program memory. In the *decode* operation, circuitry in the microprocessor *deciphers the command*, setting up all the necessary sub-circuits required to carry out the command. In the *execute* operation, the microprocessor *carries out the command*, using all the control circuitry and sub-circuits under its command. Let's watch how this plays out in the picture below.

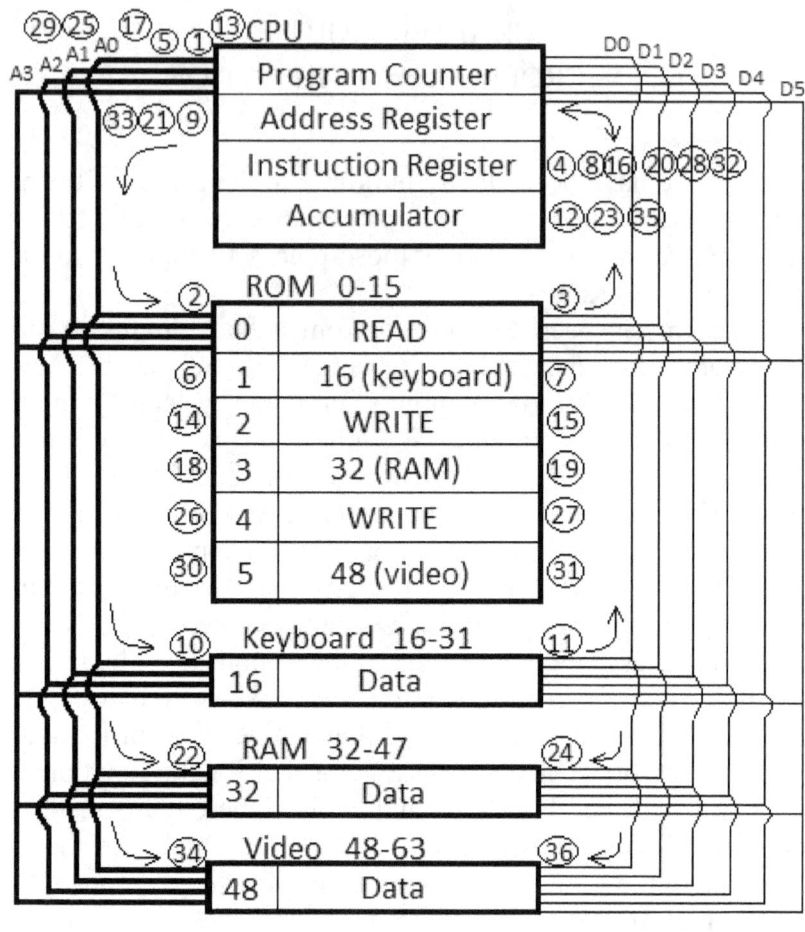

Address Bus   Data Bus

FIGURE 8-1

First, let's get comfortable with this figure through an overview. Compare this figure with Figure 7-21. The right side of Figure 8-1 is the bi-directional data bus that you already understand, from Figure 7-21. The data bus in Figure 8-1 has six bits, labelled D0 through D5. The CPU selects one of the D latches in the ROM, keyboard, RAM, or video blocks. Then, it either writes six parallel bits to it, or reads six bits parallel bits from it.

The CPU block in Figure 8-1 has more things in it. But, two of them, the instruction register and the accumulator are just CPU latches, like the one in Figure 7-21. You already understand bi-directional D latches. Everything else in the figure, below the block labelled CPU, is just D latches. I count nine, 6-bit latch groups. Six of them are clumped together in the block labelled ROM.

Still looking only at the right side of Figure 8-1, notice the arrows along the data bus. The arrow to the right of the CPU points both ways. This indicates that information flows both ways on the data bus: into and out of the CPU. The single-ended tails to the right of the ROM, keyboard, RAM, and video blocks are drawn to help you understand which way the voltage is moving on the data bus as we walk through this program's execution

Next, let's get comfortable with the devices in Figure 8-1. This is a very simple system. ROM stands for read-only memory. It's where our computer's program lives. The program commands are in the ROM, even when power is turned off. The microprocessor needs a program's commands available from the instant it is turned on, because it has to fetch, then decode, then execute. The ROM is where the microprocessor fetches from.

RAM stands for random-access memory. Unlike read-only memory, random-access memory can be both read from and written to. "Random" means the microprocessor can pick any location in RAM to read or write; it does not have to go sequentially. We will have much more to say about this selection process, shortly.

The keyboard is the location from which our microprocessor reads a computer keyboard key press. The video is the location to which our microprocessor writes the letter pressed on the keyboard, so the letter can be viewed on the LCD screen.

Now, let's look at the left side of Figure 8-1, the address bus. The address bus in this system is four bits—on four wires--labelled A0 through A3. It is part of the D latch selection process, in which the CPU chooses which of the nine D latches in the ROM, keyboard, RAM, or video it writes to or reads from. Again, we will have much more to say about this selection process, shortly.

Notice the arrows along the address bus. The address bus is not bi-directional. Voltages come out of the CPU block, either from the program counter or address register, and go to the ROM, keyboard, RAM, or video. The voltages on this bus aren't for storage, just for selection.

The address of each of the nine D latches involved in this program is drawn on the far left of each block. For example, the first ROM location is at address 0. The keyboard location is address 16. The video location is address 48. Remember that these are human, decimal numbers. The CPU will put these numbers in binary voltage form on the address bus, as part of the selection process. We will have more to say about binary numbers, shortly.

Let's get started. I will go through a simple program, in 36 steps, using Figure 8-1 as reference. During each step, look at the step number, drawn and circled in Figure 8-1, to locate where the event occurs. So that you know in advance what the program does: it reads the letter U pressed on the keyboard, stores a copy of that letter in RAM, then stores the same letter in video, for screen display.

Step 1.) You already know what step 1 will involve: a fetch. The program counter in the CPU contains a binary value of 0 at power-up. The binary 0 voltages are placed on the address bus.

Step 2.) ROM location 0 is selected.

Step 3.) The binary code for a READ command goes out of ROM and onto the data bus.

Step 4.) The READ command goes into the CPU's instruction register for decoding. The program counter bumps up by 1; it had binary 0, so now it has binary 1. The decoding of the READ command determines that the command is incomplete. More information is needed before the command execution can occur.

Step 5.) Another fetch occurs. The program counter places a binary value of 1 onto the address bus.

Step 6.) ROM location 1 is selected.

Step 7.) The binary value of decimal 16 comes out of ROM and goes onto the data bus.

Step 8.) The binary 16 enters the CPU's instruction register, and is appended to the READ command. The program counter bumps up by 1; it had binary 1, so now it has binary 2. The complete command in the instruction register is now a binary code that stands for, "READ address 16." Decoding determines that this command is complete, so command execution can now begin.

Step 9.) The binary address, 16, in the instruction register moves into the address register, then onto the address bus.

Step 10.) Keyboard location 16 is selected for reading.

Step 11.) The binary code for the letter U, the keyboard key that the user pressed, goes out of the keyboard latch and onto the data bus.

Step 12.) The binary U enters the Accumulator latch in the CPU. The Accumulator is automatically the destination of the READ command. The current execution process is complete.

Step 13.) The fetch process starts again. The program counter places a binary value of 2 onto the address bus.

Step 14.) ROM location 2 is selected.

Step 15.) The binary code for a WRITE command goes out of ROM and onto the data bus.

Step 16.) The WRITE command goes into the CPU's instruction register for decoding. The program counter bumps up by 1; it had binary 2, so now it has binary 3. The decoding of the WRITE command determines that the command is incomplete. More information is needed before the command execution can occur.

Step 17.) Another fetch occurs. The program counter places a binary value of 3 onto the address bus.

Step 18.) ROM location 3 is selected.

Step 19.) The binary value of decimal 32 comes out of ROM and goes onto the data bus.

Step 20.) The binary 32 enters the CPU's instruction register, appended to the WRITE command. The program counter bumps up by 1; it had binary 3, so now it has binary 4. The complete command in the instruction register is now a binary code that stands for, "WRITE address 32." Decoding determines that this command is complete, so command execution can now begin.

Step 21.) The binary address, 32, in the instruction register moves into the address register, then onto the address bus.

Step 22.) RAM location 32 is selected for writing.

Step 23.) The contents of the CPU's Accumulator latch is placed on the data bus. The Accumulator is automatically the source for the WRITE command. Remember, it contains the binary code for the letter U, acquired in step 12.

Step 24.) The binary U value enters RAM location 32 for storage. The current execution process is complete.

Step 25.) The fetch process starts again. The program counter places a binary value of 4 onto the address bus.

Step 26.) ROM location 4 is selected.

Step 27.) The binary code for a WRITE command goes out of ROM and onto the data bus.

Step 28.) The WRITE command goes into the CPU's instruction

register for decoding. The program counter bumps up by 1; it had binary 4, so now it has binary 5. The decoding of the WRITE command determines that the command is incomplete. More information is needed before the command execution can occur.

Step 29.) Another fetch occurs. The program counter places a binary value of 5 onto the address bus.

Step 30.) ROM location 5 is selected.

Step 31.) The binary value of decimal 48 comes out of ROM and goes onto the data bus.

Step 32.) The binary 48 enters the CPU's instruction register, appended to the WRITE command. The program counter bumps up by 1; it had binary 5, so now it has binary 6. The complete command in the instruction register is now a binary code that stands for, "WRITE address 48." Decoding determines that this command is complete, so command execution can now begin.

Step 33.) The binary address, 48, in the instruction register moves into the address register, then onto the address bus.

Step 34.) Video location 48 is selected for writing.

Step 35.) The contents of the CPU's Accumulator latch is placed on the data bus. The Accumulator is automatically the source for WRITE commands. Remember, it contains the binary code for the letter U, acquired in step 12.

Step 36.) The binary U value enters video location 48 for storage and viewing on a screen. The current execution process is complete. Our program is complete.

That's all there is to it. Our simple program, for our simple microprocessor system, is done. What does our microprocessor do next? You know that it has to fetch a command. Our next command, at ROM address 6 could be, GOTO 0. The GOTO command replaces the current contents of the program counter with the value in the second part of the GOTO command. Since our command is GOTO 0, a binary 0 would be placed in the program counter, so that the fetch would be from ROM address location 0. This effectively starts our program again, from the beginning. In fact, our program restarts every time it finishes, due to the GOTO 0 command at the end. We call this an infinite loop. The program loops back from end to beginning every time, forever, or until the power supply to the microprocessor is turned off.

CHAPTER NINE
BINARY NUMBERS

Now we can explore the details of the address bus. I intentionally worked backwards, here. I could have shown you the nuts and bolts of how address selection works first. I preferred to first show an address bus in action, so that you could see why we need one. And, so you can appreciate its importance.

Binary groups can represent whatever we want them to represent: command code, numbers, alphanumeric characters, etc. On the address bus, we treat the binary voltages as numbers: as numerical addresses, just like our houses have address numbers. You need to know a little about how we represent numbers in binary.

We construct the binary number system the same way we construct the decimal number system. In the decimal number system, we have ten distinct digits. In binary, we have two distinct digits: 0 and 1. To prevent confusion when talking about binary and decimal, I will append a b at the end of a binary number, and a d at the end of a decimal number. So, if I have 0b asterisks, I don't have any asterisks. If I have 1b asterisks, I have this many asterisks: *.

How do I represent ** asterisks? Just like in decimal, when I want to get to the next number, I add 1b to the current number. This is called counting. Counting is how I went from 0b to 1b: I added 1b to 0b to obtain 1b. Now I want to add 1b to 1b to get to the next number, which will represent ** asterisks.

But I have run out of numbers. There is no one-digit number after 1b, in binary. What did I do in decimal, when I was counting (adding 1 to the current number) and ran out of numbers? This happened at 9d. After 9d, there is no higher one-digit number in decimal. When I reached 9d and added 1d to 9d, I reset the 9d back to 0d and added a 1d to the next column to the left. My next number, after 9d, became 10d. See the example, below:

```
   1d<----------(CARRY COLUMN)
      9d<----(STARTING COUNT)
 +    1d<----(ADD 1d)
   ────────
   1  0 d<---(NEW COUNT: 10d)
```

TABLE 9-1

What is the significance of the next column to the left? Numbers in the new column, second from the right, are worth ten times as much as numbers in the first column (in decimal). So, a 1d in the first column is worth * asterisk. But 10d has a 1d in the second column from the right, so it is worth ten times as many asterisks: **********. In fact, every time we move another column to the left, the value of that decimal digit increases another ten-fold.

Following the same rules we used for decimal, we find that the binary number after 1b is 10b. How did we arrive at this answer? We reset the 1b to 0b, and add a 1b to the next column. The new number, 10b, represents ** asterisks. See the example, below:

```
   1b<---------(CARRY COLUMN)
      1b<----(STARTING COUNT)
 +    1b<----(ADD 1b)
   ────────
   1  0b<----(NEW COUNT: 10b)
```

TABLE 9-2

Now you can see the importance of using a b or d after our numbers. In decimal, 10d represents ********** asterisks. In binary, 10b represents ** asterisks. Without the b or d, they look the same: 10.

In decimal, a number in any column is worth 10d times as many asterisks as the same number in the column to its right. So, in decimal 321d, the 3d is really worth (3d)x(100d), or 300d; the 2d is really worth (2d)x(200d), or 200d; and the 1d is really worth (1d)x(1d), or 1d.

Add them up, and we write it as 321d.

In binary, a number in any column is worth 10b times as many asterisks as the same number in the previous column, to its right. 10b in binary is 2d in decimal, to us humans, comfortable speaking in decimal. So, speaking as a human in decimal, a binary number in any column is worth double, or twice as many asterisks as the same number in the column to its right.

If we continue counting in binary, we keep using the rule of resetting to 0b and adding 1b to the next column to the left. For example, if we have counted to 11b, the next number is 11b + 1b. We add 1b to the rightmost column of 11b. We have run out of digits, so we reset the resultant to 0b and carry the 1b to the next column to the left. The carry 1b plus the 1b in the second column also overflows, so we reset to 0b again, and add 1b to the next left column. The answer becomes 11b + 1b = 100b.

So, counting in binary looks like this:

BINARY	DECIMAL
000000b	0d
000001b	1d
000010b	2d
000011b	3d
000100b	4d
000101b	5d
000110b	6d
000111b	7d
001000b	8d
001001b	9d
001010b	10d

TABLE 9-3

We say that a binary number in any column is worth double, or twice as many asterisks as the same number in the column to its right. Using this knowledge, we can convert any binary number to decimal by using the chart on the following line:

128   64   32   16   8   4   2   1

TABLE 9-4

The above sequence is the weighting value of the bit position for each bit of an 8-bit binary number. Each value (after the rightmost) is double the number to its right. To convert 00111011b from binary to decimal, line up the 8 binary digits under their corresponding weighting values. Multiply vertically, then add the multiplication results horizontally. See the example below.

128	64	32	16	8	4	2	1	(DECIMAL)
x 0	x 0	x 1	x 1	x 1	x 0	x 1	x 1	(BINARY)

$$0 + 0 + 32 + 16 + 8 + 0 + 2 + 1 = 59 \text{(DECIMAL)}$$

TABLE 9-5

59d is the decimal form of the binary number 00111011b.

There are formal ways to convert the other way, from decimal to binary, but they involve long division. Most students nowadays prefer a trial and error solution (or use a scientific calculator). It's like the above method, but in reverse. For example, to convert 50d to binary, prepare this chart:

128	64	32	16	8	4	2	1	(DECIMAL)
x 0	x 0	x 0	x 0	x 0	x 0	x 0	x 0	(BINARY)

$$0 + 0 + 0 + 0 + 0 + 0 + 0 + 0 = 0 \text{ (DECIMAL)}$$

TABLE 9-6

Start from the left, and work toward the right. Find the biggest number that does not exceed 50d. That would be 32d. Replace the 0b under the 32d with a 1b. Keep a separate running tally of the numbers above the bits we change from 0b to 1b. Our tally now only contain 32d. (See below.)

128	64	32	16	8	4	2	1	(DECIMAL)
x 0	x 0	x 1	x 0	x 0	x 0	x 0	x 0	(BINARY)
0 +	0 +	32 +	0 +	0 +	0 +	0 +	0	= 32 (DECIMAL)

TABLE 9-7

Keep working your way to the right, finding the next decimal number we can add to 32d, without the sum exceeding 50d. 16d is the next number, because 32d+16d=48d does not exceed 50d. Change the 0b under the 16d to a 1b, and add the 16d to our running tally: 32d+16d=48d. Our chart now looks like this:

128	64	32	16	8	4	2	1	(DECIMAL)
x 0	x 0	x 1	x 1	x 0	x 0	x 0	x 0	(BINARY)
0 +	0 +	32 +	16 +	0 +	0 +	0 +	0	= 48 (DECIMAL)

TABLE 9-8

We repeat this process, stopping once our running tall reaches 50d. We are at 48d. Obviously, we must not add 8d, because 48d + 8d = 56d, which exceeds 50d. In fact, at 48d, we only need 2d more to exactly arrive at 50d. So, we can also skip over 4d. We change the 0b under the 2d to a 1b, and add the 2d to our running tally: 48d + 2d = 50d. Our chart now looks like this:

128	64	32	16	8	4	2	1	(DECIMAL)
x 0	x 0	x 1	x 1	x 0	x 0	x 1	x 0	(BINARY)

$$0 + 0 + 32 + 16 + 0 + 0 + 2 + 0 = 50 \text{ (DECIMAL)}$$

TABLE 9-9

So, we started with decimal 50d and wanted to know the binary equivalent. We conclude that 00110010b is the binary equivalent of 50d. The first two zeroes on the left are not required. The answer doesn't change if you were to write 110010b, instead of 00110010b.

CHAPTER TEN
THE ADDRESS BUS AND DECODER

Returning to our address bus, assume we have a 4-bit address bus. Assume we have labelled the four wires of the bus as A3, A2, A1, and A0. We agree that the wire labelled A3 stands for the most significant bit, and the one labelled A0 stands for the least significant bit. In other words, that they follow the chart below:

A3	A2	A1	A0	(WIRE LABEL)
8	4	2	1	(BIT VALUE OR WEIGHTING)

Let's say we want to enable the latch that has the decimal address of 9d. What binary address do we put on the data bus? Well, 9d in decimal is the same number as 1001b in binary. They both describe ********* asterisks. Since we put voltages on the address bus, our address bus would physically contain the following voltages at the following locations:

A3	A2	A1	A0	(WIRE LABEL)
8	4	2	1	(BIT VALUE OR WEIGHTING)
1	0	0	1	(BINARY VALUE)
5v	0v	0v	5v	(VOLTAGE)

Now that you understand what is physically on the address bus, you next need to understand how these voltages are decoded: how they select only one of the many 6-bit data latches in Figure 8-1. Go back, right now, to Figure 7-12. You see, in Figure 7-12, a box labelled C with the ability to select one of several data latches: to make the voltage at the E (enable) input of one data latch 5 volts, while keeping all the other E inputs of the non-selected data latches 0 volts. This section explains what is inside that mysterious box labeled C: the *address decoder*.

For the moment, let's simplify our Figure 8-1 even further. Let's pretend that ROM is only one latch group, RAM is only one latch group, the keyboard is only one latch group, and video is only one latch group. Four latch groups require four unique addresses (2^2=2x2=4) to decode them. The addresses would be: 00b, 01b, 10b, and 11b in binary, for ROM, keyboard, RAM, and video, respectively. Let's label the address lines A5 and A4. See the picture below to learn how the microprocessor system decodes A5 and A4:

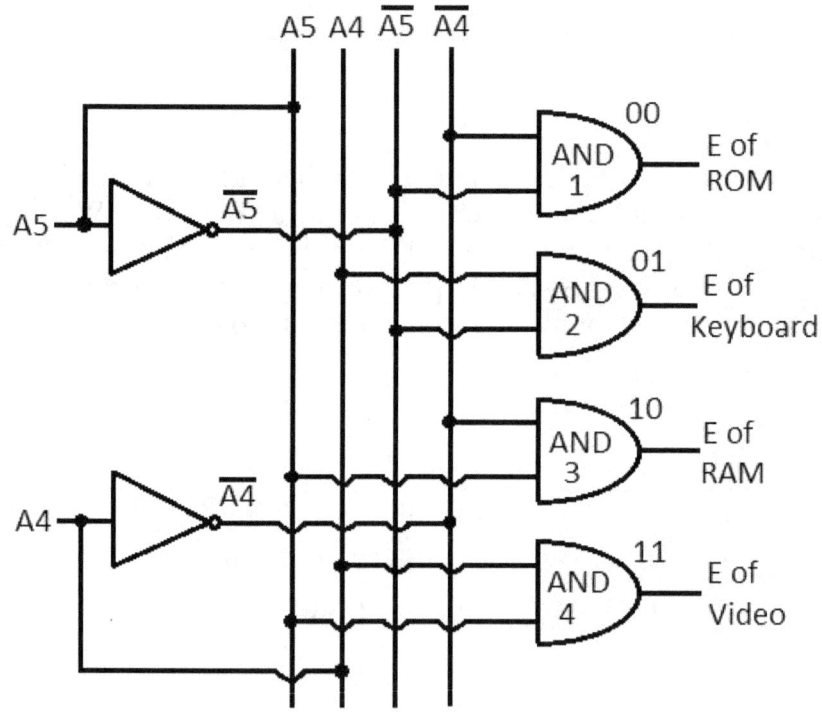

FIGURE 10-1

In Figure 10-1, a small microprocessor address decoder, I have arranged the middle section purely for ease of understanding. This internal bus doesn't lead anywhere out of the picture. There only two inputs, A5 and A4. There are four outputs: E of ROM, E of keyboard, E of RAM, and E of video. E stands for enable. Remember, each D latch is accessible only when its E input is high. The logic table (sometimes called the truth table) that completely describes the behavior of this address decoder is:

Address Decoder
Truth Table

A Inputs		E Outputs			
A5	A4	ROM	Key	RAM	Vid
0	0	1	0	0	0
0	1	0	1	0	0
1	0	0	0	1	0
1	1	0	0	0	1

FIGURE 10-2

Reading the truth table one row at a time, we find summarized what we described in the paragraphs above. For example, in the first row, when A5 and A4 are both low, only the E output to ROM is high, meaning only the ROM latch is enabled.

Let's look at Figure 10-1 to see how this decoding takes place. A5 and A4, generated by the CPU, enter the address decoder. They are immediately inverted by inverters, so that their logical opposites are available as NOT A5 and NOT A4. In Figure 10-1, these inversions are shown as A5 and A4 with lines above them. It's hard for word processing software to put lines above characters, so instead we put the word, "NOT" in front of the inverted logic. Another technique is to put a forward slash in front, as in /A5 and /A4.

Now, four different logic states are available to the four AND gates. Non-inverted versions of A5 and A4 go directly to AND4. So, per the truth table of AND gates, AND4's output goes high when A4 and A5 are both high. This creates row four of the truth table.

AND1's two inputs both come from the inverters. So, when A5 and A4 are both low, inverter outputs /A5 and /A4 are both high, causing AND1's output to go high. This creates row one of the truth table.

AND2's output goes high (true) when A4--which doesn't get inverted--is high and A5--which does gets inverted--is low. This creates row two of the truth table.

AND3's output goes high (true) when A5--which doesn't get inverted--is high and A4--which does gets inverted--is low. This creates row three of the truth table.

The creation of complexity from simplicity follows the following process in digital electronics. We start with voltage sources, transistors, and resistors. We put them together to make inverters. We put inverters together to make gates. We put gates together to make latches. We put gates and latches together to make more complex circuits. We put more complex circuits together to make machines and devices like microprocessors.

The circuit in Figure 10-1 is an example of putting together basic building blocks--just gates in this case, no latches--to create a slightly more complex circuit. Its name is a 2-to-4 binary decoder. As we create these slightly more complex circuits, we give them names, and encapsulate them in new logic diagrams. This way, we don't have to draw the circuitry in Figure 10-1 repeatedly. We can instead draw the much simpler logic diagram for the 2-to-4 binary decoder shown below:

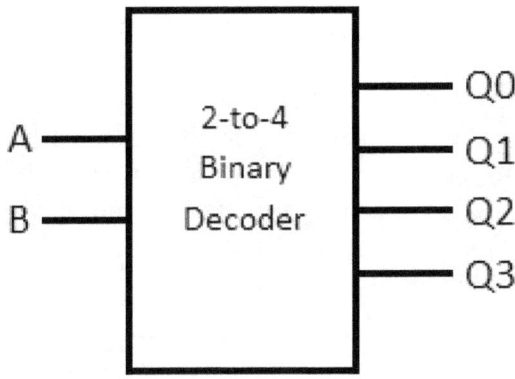

FIGURE 10-3

The input and output labels above are generic ones you will find in documentation for decoders. The logic diagram in Figure 10-3 focuses our attention on the inputs on the left and the outputs on the right. Along with the truth table in Figure 10-2 and your understanding of how the decoder works, you no longer have to be concerned with the circuitry it encapsulates in Figure 10-1. Also, you don't have to concern yourself with the resistors and transistors that Figure 10-1's AND gates and inverters encapsulate.

Let's return, now, to Figure 8-1: our microprocessor system. In the above analysis, I simplified the address bus by making each addressable subsystem contain only one latch: one ROM latch, one RAM latch, one keyboard latch, and one video latch. In a real system, each addressable subsystem contains many latches. In our system of Figure 8-1, each subsystem has 16 latches. The CPU accessed 6 of the 16 ROM latches. Our program accessed only one latch from each of the RAM, keyboard, and video subsystem, but it could have accessed 16 latches in each.

A microprocessor-based system, then, actually performs latch selection in a 2-step process. Did you notice that in our simplified analysis in Figure 10-1 I labelled the address bus wires as A5 and A4? And yet, in our actual circuit of Figure 8-1, I labelled the address bus wires A3, A2, A1, and A0. This is evidence of a 2-step process.

Our system's finished address bus really requires 6 address lines. In a microprocessor system, the highest order bits are separated from the lower order bits; the highest order bits are then sent to an address decoder. Thus, the highest order bits are used to generate the enable signal that picks just one subsystem.

The lower order bits are then used to select just one latch of the many contained in the one active subsystem. Though the lower order bits go to all the subsystems, they are ignored by the non-selected subsystems. So, let's apply this new-found knowledge to our system. Let's remove the parts of Figure 8-1 we don't need—like the data bus-- and insert what we have just learned. See the picture below:

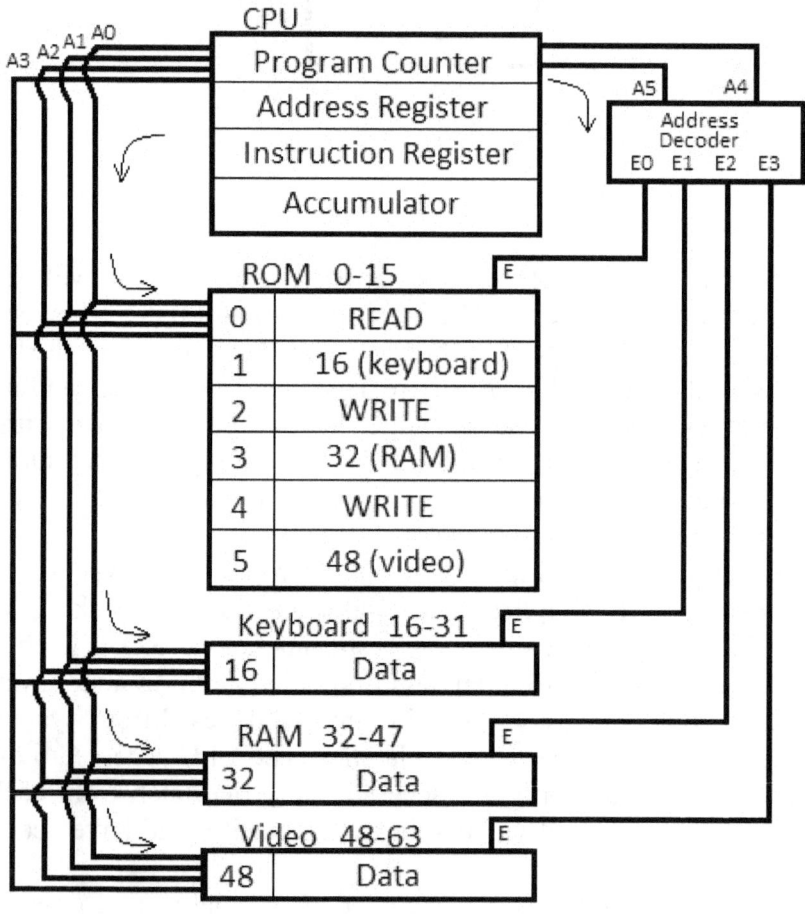

FIGURE 10-4

In Figure 10-4, you can see how the 2-step address selection process plays out in our system. Now you can see all six of the address bits our CPU generates. You can see the separate path the higher-order address bits, A5 and A4, take to the address decoder. This is step one. The address decoder is the 2-to-4 decoder from Figure 10-3. The address decoder activates only one of the four subsystems, by raising its E wire high.

Once the chosen subsystem is activated by the address decoder, address lines A3, A2, A1 and A0 are further decoded *inside the subsystem* to pick the one latch group of 16 to write to or read from. This is step two.

From the CPU's point of view, it puts 64 different addresses on its address bus. That's the number of unique combinations of a 6-bit binary number (the address bus is 6 bits wide). The CPU doesn't know about the address decoder or about four different subsystems. From the subsystems' (ROM, RAM, keyboard, and video) point of view, each one of them only reacts to 16 different addresses: decimal 0d to 15d, which is the same as binary 0000b to 1111b. They also know nothing about the address decoder or about other subsystems. They know nothing about any addresses higher than 15d or 1111b.

It's the address decoder, or perhaps we should say the designer of the address decoder, that creates the memory map you see in Figure 10-4. Look closely, and you will see that the ROM responds to addresses 0d to15d, the keyboard responds to addresses 16d to 31d, the RAM responds to addresses 32d to 47d, and the video responds to addresses 48d to 63d, that are sent out of the CPU. Look at the chart below:

DEVICE	A5	A4	A3 through A0	A5 through A0	A5-A0 Decimal
ROM	0b	0b	0000b-1111b	000000b-001111b	0d-15d
KEY-BOARD	0b	1b	0000b-1111b	010000b-011111b	15d-31d
RAM	1b	0b	0000b-1111b	100000b-101111b	32d-47d
VIDEO	1b	1b	0000b-1111b	110000b-111111b	48d-63d

FIGURE 10-5

Each row in Figure 10-5 shows the address range for one subsystem, and how it is derived. The column with the heading, "A5-A0 decimal" shows the addresses that the CPU sends out onto its 6-bit address bus, and what devices respond to them. The decimal numbers don't make much sense until you look at them in binary, as shown in the adjacent column with the heading, "A5 through A0."

For example, the chart says that the VIDEO subsystem responds to the decimal range of addresses 48d through 63d, which is the binary range of addresses 110000b through 111111b. (You already know how to do decimal to binary and binary to decimal conversions.)

But, where do these binary numbers come from? Well, the column labelled, "A3 through A0," shows what each subsystem sees at its address input wires. Clearly, each subsystem sees the same address range: 0000b through 1111b.

The columns labeled A5 and A4 complete the story. Address bus voltages A5 and A4--via the address decoder--select a device in one row of the chart, to access that device's internal 0000b-1111b addresses. So, we now know where the VIDEO subsystems address range of 110000b through 111111b comes from. We use numbers from the VIDEO row in Figure 10-5. We write, from left to right, the binary values for: A5 (a 1b), then A4 (also a 1b), then the 4-bit binary values for A3 through A0 (0000b through 1111b). We say we "append" the values, 0000b through 1111b, to 11b, to produce 110000b through 111111b. These are the address numbers the computer engineer who designed this system uses to access VIDEO.

As stated above, once the chosen subsystem is activated by the address decoder, address lines A3, A2, A1 and A0 are further decoded *inside the subsystem* to pick the one latch group of 16 to write to or read from. Now, I'll show you how the decoding inside the RAM, inside the video, etc., happens.

Actually, the address decoding inside the subsystem is very similar to the address decoding you already have witnessed outside the subsystem, as in Figure 10-4. The difference is, in Figure 10-4 we had to select one of four addresses, whereas inside each subsystem, we must select one of sixteen addresses. We have four address lines, A3-A0, to select one of those sixteen addresses, so we need a 4-to-16 decoder.

We could design a new, 4-to-16 binary decoder. You may be able to design it yourself, from Figure 10-1. (Hint: use four inverters and sixteen, 4-input AND gates. Instead, let's follow our procedure of building new, progressively more complex circuits out of our old ones. Let's build our 4-to-16 decoder out of two 2-to-4 decoders and some extra AND gates. We'll even save a few transistors doing it this way. See the picture below:

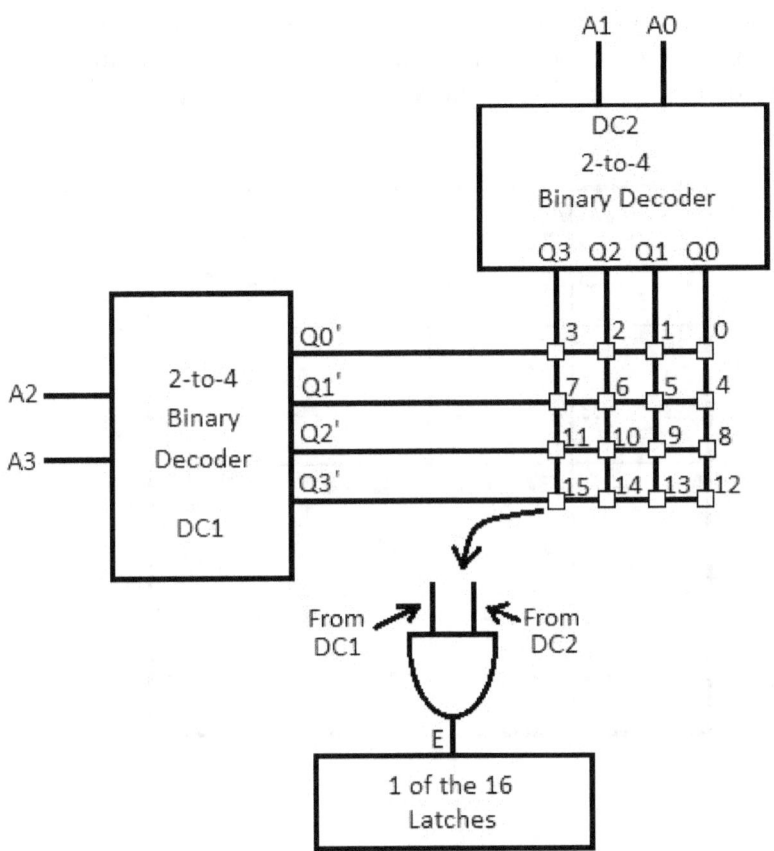

FIGURE 10-6

The circuit shown in Figure 10-6 is inside each of the four subsystems, ROM, keyboard, RAM, and video. DC1 and DC2 are two of our 2-to-4 binary decoders. They connect to the world outside them via lines A3, A2, A1, and A0, as shown above.

The eight wires, Q0 through Q3 and Q0' through Q0', never

intersect: never touch each other in any way. They create a 4x4 grid. At the intersection of each wire pair, I have drawn a small box. I have zoomed in on the contents of box 15. The contents of box 15 are displayed at the bottom of Figure 10-6: an AND gate and one of the sixteen latches. This is what you will find in each of the 16 small boxes. I have also placed a decimal number next to each small box. This number is the decimal equivalent of the binary number on address wires A3 through A0 that activates the latch inside the small box. See the picture below, as we discuss how this circuit works.

A3A2 A1A0 0 0 1 1 Q0' Q3 3	A3A2 A1A0 0 0 1 0 Q0' Q2 2	A3A2 A1A0 0 0 0 1 Q0' Q1 1	A3A2 A1A0 0 0 0 0 Q0' Q0 0
A3A2 A1A0 0 1 1 1 Q1' Q3 7	A3A2 A1A0 0 1 1 0 Q1' Q2 6	A3A2 A1A0 0 1 0 1 Q1' Q1 5	A3A2 A1A0 0 1 0 0 Q1' Q0 4
A3A2 A1A0 1 0 1 1 Q2' Q3 11	A3A2 A1A0 1 0 1 0 Q2' Q2 10	A3A2 A1A0 1 0 0 1 Q2' Q1 9	A3A2 A1A0 1 0 0 0 Q2' Q0 8
A3A2 A1A0 1 1 1 1 Q3' Q3 15	A3A2 A1A0 1 1 1 0 Q3' Q2 14	A3A2 A1A0 1 1 0 1 Q3' Q1 13	A3A2 A1A0 1 1 0 0 Q3' Q0 12

FIGURE 10-7

If Figure 10-7 scares you, let's first analyze the circuit in Figure 10-6 without it. Address lines A3 and A2 enable one of the four rows on the output of DC1. This means that only one of the rows has a high voltage on it. This also means that only the four AND gates in that row have a high on one input. So, for example, if A3 and A2 are both low, only row Q0' goes high, so only the top row's AND gates have one input high.

But, to activate a latch, we require both a row and a column to be enabled. Address lines A1 and A0 enable one of the four columns on the output of DC2. This means that only one of the columns has a high voltage on it. This also means that only the four AND gates in that column have a high on one input. So, for example, if A1 and A0 are both low, only column Q0 goes high, so only the right column's AND gates have one input high.

However, for the output of a 2-input AND gate to be high, both of its inputs must be high. If you follow through the example in the previous two paragraphs, you will find that only the AND gate in the upper right-hand corner has both of its inputs high. It is the only AND gate whose inputs are driven by both the top row's (Q0') and the right column's (Q0) high outputs. So, when A3 through A0 are 0000b, only the AND gate hidden in the small box labelled 0 in Figure 10-6 has a high output. This high output pulls high the enable input of its latch, for reading or writing to that latch.

Now look at Figure 10-7. It is laid out in the same order as the grid in Figure 10-6. The decimal number at the bottom of each square in Figure 10-7 is the same as the one in each corresponding little box in Figure 10-6. It's the decimal form of the address on A3 through A0 that activates the latch inside the little box of Figure 10-6.

To continue, let's look at just one square in Figure 10-7. Look at the square in the lower, right-hand corner: the one with a 12 in it. You already know that 12d is the decimal form of the address on A3 through A0 that activates the latch inside the little box of Figure 10-6. The second row of the square is 1100b, the binary form of 12d. 1100b describes the actual voltage levels on A3 through A0. Directly above each digit in 1100b, you will find the corresponding A3, A2, A1, or A0 wire that drives that voltage level. You will find that 1100b is split in half, and displayed as 11b then a space, then 00b. That's because A3 and A2 generate a Q' output, and A1 and A0 generate a Q output, through their corresponding binary decoders. In this specific case, the 11b sends Q3' high, and the 00b sends Q0 high. These two highs meet at the AND gate that activates latch number 12d.

CHAPTER ELEVEN
THE CONTROL BUS

At this point, we have covered the devices and program of our simple microprocessor system, in Figure 8-1; the external address decoder involved in device selection, in Figure 10-4; and the internal address decoder, hidden in each device, in Figure 10-6. You have learned how the address and data bus work. There's another bus, involved in all this. It's called the control bus. Once again, I'll hide the previous connections from Figures 8-1 and 10-4 that might confuse you. See the picture below:

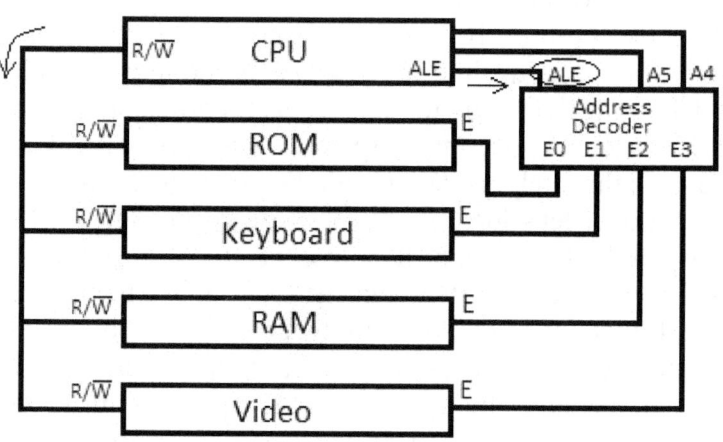

FIGURE 11-1

The new items are the ALE and R/W outputs from the CPU. They make up the simple control bus for our simple system. The name, "control bus," is self-explanatory. The control bus is the signals that the CPU uses to control communication with subsystems.

The new items here are not really new to you. You already know, from Figure 7-22, that bi-directional latches require an R/W input to control whether data is being input to or output from them. The CPU provides this voltage. Remember, a high on R/W means "read." The CPU reads the latch, so data comes out of the latch. A low on R/W means "write." The CPU writes to the latch, so data goes into the latch. In Figure 11-1, I drew R/W control into all four subsystems. If any of them are unidirectional, they don't need an R/W line.

You also know from Figure 7-22 that latches require a high on their E line (enable line) to turn them on. We have already covered that control process above. However, if you look back at the address decoder we designed in Figure 10-1, you will see that there is no way to turn all four outputs off, simultaneously. One of its outputs has to be high: either E0, E1, E2, or E3. This, in turn, means that one of the four subsystems has to be on. This situation is a problem. No subsystem should be on, unless the CPU is deliberately accessing it. Data corruption can result.

Our CPU provides a control line to help overcome this problem. The line is named "ALE," for "address latch enable." The CPU pulls this line high when it has put an address on the address bus. Only when the ALE line is high should one of the subsystems be on. Only when the ALE line is high should one of the address decoder outputs be high. When the ALE line is low, all four address decoder outputs should be low.

In Figure 11-1, you will see that the ALE line goes to the address decoder. The address decoder of Figure 10-1 obviously needs a minor change. The four AND gates need to become 3-input AND gates. All of the new, third inputs to the AND's should be connected together, and then connected to the ALE signal. The picture below reflects these changes in our address decoder:

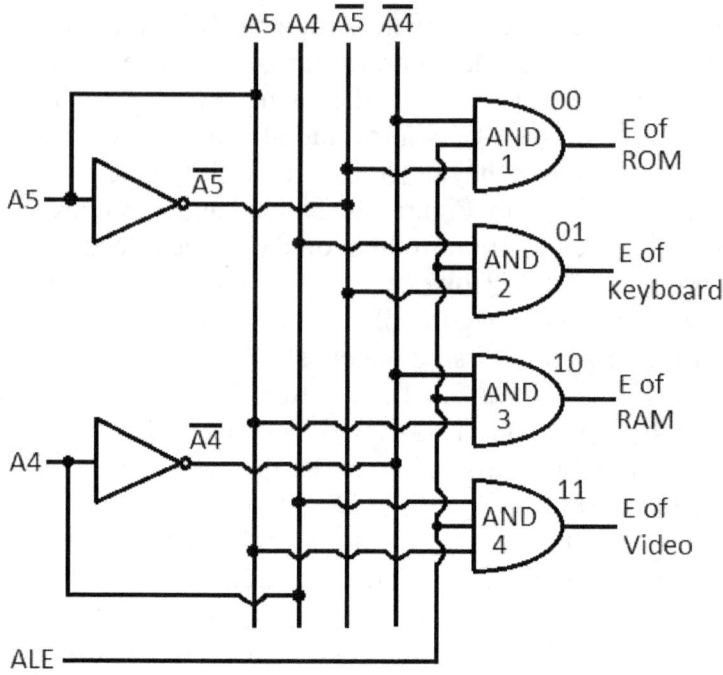

FIGURE 11-2

Different microprocessor manufacturers have slightly different ways of implementing a control bus. For example, some microprocessors split our R/W line into two separate lines: an R line for reading, and a W line for writing. Some microprocessors have one Enable (E) line for memory (RAM, ROM) and a separate line for input/output (I/O) devices (keyboard, video). Interrupt and DMA control, if implemented, require control bus lines.

Whatever the design, proper timing of the arrival of information on the address and data buses is critical. If a microprocessor only contains the CPU, but not the subsystems (peripherals), that microprocessor's manufacturer will often sell peripherals designed specifically to work with the microprocessor and its control bus. This helps to prevent timing bugs, and makes life easier for designers.

CHAPTER TWELVE

THE SYSTEM CLOCK

Now, let's look at some other secrets that make the CPU work. First, let's consider the sequential nature of the system. I said that the microprocessor does three things, over and over: fetch, decode, and execute. This implies an ordered sequence of events. First this happens; then this happens, later in time. None of the gates or latches we have studied so far change over time.

Because of this requirement, all microprocessor systems have a clock. This is not the kind of clock that tells the time of day. Rather, it is a square-wave oscillator. It is a device that creates a digital voltage that is high for a brief period of time, then low for a brief period of time, then high for a brief period of time, etc. It never stops this ceaseless repetition, until the power supply is turned off. If we were to plot a graph, with voltage on the vertical axis, and time on the horizontal axis, it would look like this:

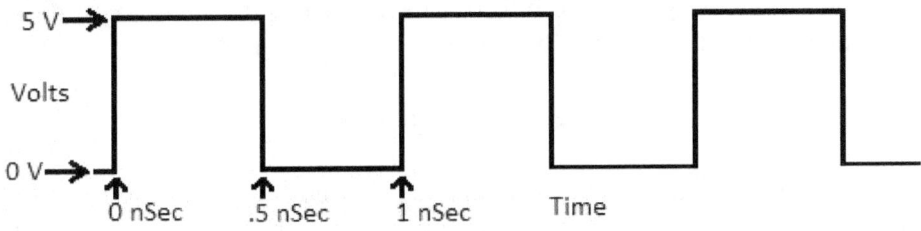

FIGURE 12-1

The clock pictured in Figure 12-1 is high for .5 nanoseconds, then low for .5 nanoseconds, then repeats. A full cycle is a high followed by a low, so a full cycle takes 1 nanosecond. A nanosecond is one billionth of a second. In one second, this clock repeats the cycle a billion times. In electronics, we like to say "mega-," instead of million, and "giga-," instead of billion. So, this figure describes a 1 gigahertz (1 billion cycles per second) clock. We refer to 1 gigahertz as the clock's frequency. Think of frequency as how frequently the clock repeats its cycle in one second.

How does one build a clock like this? If we want to build upon circuits we have already studied, then we may build one like the one shown in the following picture:

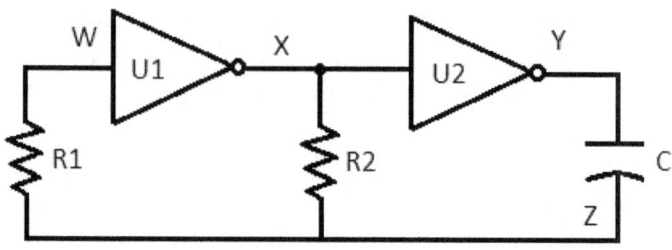

FIGURE 12-2

The circuit shown above will oscillate in a square-wave pattern. Its frequency can be calculated with the following equation:

f=1/(2.3xR2xC)   [where R1 is chosen to be 10 x R2]

We are very familiar with inverters (U1 and U2) and with resistors (R1 and R2). This circuit introduces a new component: the capacitor. The capacitor is on the far right, labelled C. Its circuit diagram is the straight line and the curved line. (Sometimes, it is represented as two parallel lines.)

Let's figure out why this circuit oscillates. We'll start by looking at the left side only. See the picture below:

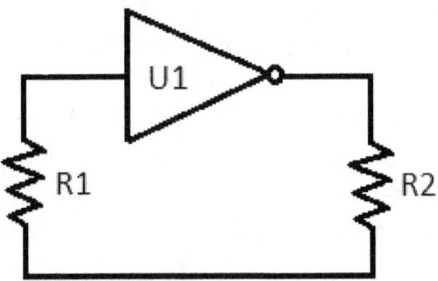

FIGURE 12-3

Think about this circuit, for a while. Have you noticed something strange? Assume the input is high. That means, as an inverter, the output must be low. But the low output feeds back to the input. But, we just said the input was high. How can the input be both low and high? Does the low output turn the high input to low, causing the low output to immediately go high? But then, does the new high output go back around to replace the old low input, causing the output to invert back low? Does this go on forever? Is this like Krypto, Superman's super-dog, chasing its tail at lightning speed?

Yes, this is an out-of-control oscillator, if it ever oscillates at all. What is needed is something to slow it down, to get it in control. Whenever we need something in electronics to slow things down, we turn to capacitors. Capacitors introduce the element of time into electronics. Inductors do, too. But inductors also have the nasty habit of causing sparks. Inductors are good for causing sparks across your car's spark plugs, but are bad for most electronic circuits. So, capacitors are more popular.

Let's look at Figure 12-2 at the exact moment when point W switches to 5 volts, point X switches to 0 volts, and point Y switches to 5 volts. These opposite relationships have to hold, due to the inverting nature of inverters. What's to prevent the 0 volts at X from feeding back immediately to W, and causing the whole nasty runaway problem mentioned above?

The voltage at point Z prevents this problem. The instant before this new state, point Y was in its old state, 0 volts. Point Z was at 5 volts; it had to be at 5 volts to give point W its 5 volts, to initiate the current state.

But, at the moment the new state occurs, when the voltage of point Y jumps to 5 volts, the voltage at point Z jumps to 10 volts! Voltages across capacitors take time to change; we will explore this later. At the end of the old state, there was a 5 volt difference between the top and bottom of the capacitor, C. The top, point Y, was 0 volts; the bottom, point Z, was 5 volts. A capacitor's potential difference can't change immediately: it takes time to change. So, if the top of the capacitor gets instantly raised up by positive 5 volts, then the bottom of the capacitor gets raised up by positive 5 volts. The bottom of the capacitor instantly jumps from 5 volts to 10 volts. Though voltages at both ends of the capacitor change instantly, the potential difference does not change instantly. At the instant of state change, the voltage difference, end to end, on the capacitor is kept the same as the instant before the state change.

So, the 0 volts at point X is blocked from getting to point W by the 10 volts at point Z. But, how does the circuit oscillate to the reverse state? The capacitor causes this, too. Voltages across capacitors don't change instantly, but they do change eventually. Over time, the voltage at the point Z end of the capacitor settles to the voltage level point X wants, 0 volts. When it arrives at 0 volts, that voltage, applied to point W, initiates a new change of state, so that W=0 volts, X=5 volts, and Y=0 volts.

Let's trace the next sequence of events. Let's call the instant before the new state T-, and the instant after the new state, T+. At T-, there was a 5 volts difference in voltage from one end of the capacitor to the other. At T-, point Y was at 5 volts, and point Z was at 0 volts. At T+, point Y drops down by 5 volts, to 0 volts. So, the voltage at point Z has to drop down by 5 volts, too. It drops from 0 volts to -5 volts!

Of course, the - 5 volts at point Z is more than enough to prevent the 5 volts at point X from reaching point W, to start the runaway oscillation. Over time, however, the -5 volts at point Z gradually reaches the voltage level sought by point X, 5 volts. After this length of time, that 5 volts reaches point W, to start the next state, which is where we began this discussion.

CHAPTER THIRTEEN
CAPACITORS

Once again I have chosen to work backwards in my presentation. I have chosen, above, to show the capacitor in action, before explaining the details of how it works. Hopefully, it has intrigued you, and whetted your appetite to learn more about how this amazing device works, as described below.

If, however, you are impatient to learn about how microprocessors work, you can skip over the next section on capacitors. Missing out on it should not prevent you from understanding how microprocessors work. You can always come back to this section later, and treat it as a bonus learning session, as Appendix D. I have included this section on capacitors so you can better understand how clock circuits work, because a microprocessor has to have a clock. Understanding how capacitors work will also help you to understand how crystal-based oscillators work, in the next chapter.

Capacitors are very important components in electronics circuits. Without them, we would have no electronic communication systems such as radio, television, and Wi-Fi. They are used to create and receive transmitted signals, selectively tune them in, filter them for treble and bass. They protect our high speed digital circuits, like microprocessors, from unwanted noise (radio frequency interference [RFI]). It behooves us to give them some attention.

To get a feeling of how capacitors work, let's return to our water faucet pressure analogy. Look at the picture below:

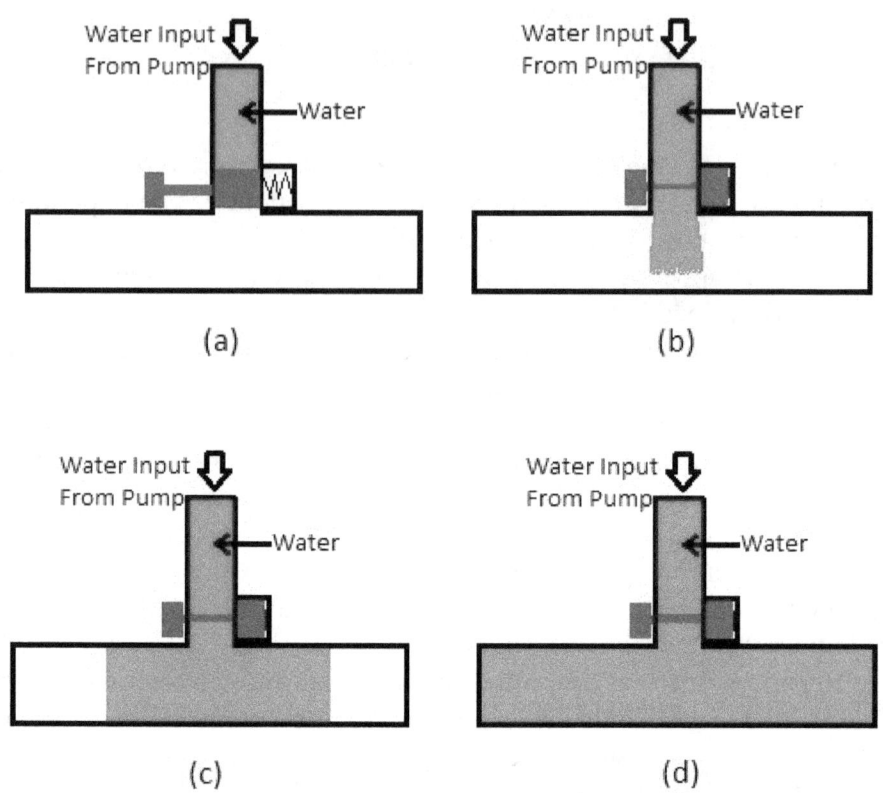

Water Input From Pump
Water

(a)

Water Input From Pump
Water

(b)

Water Input From Pump
Water

(c)

Water Input From Pump
Water

(d)

FIGURE 13-1

We said earlier that a capacitor slows down the change in voltage across it. The simplest analogy would be in the picture above. I have changed our water system so that the valve opens into a very large, closed vat. Imagine the vat being as deep into the page as it is wide across the page. In part (a), the valve is closed. All the pressure from the pump, transmitted through the water, presses against the closed valve. Let's call the pump pressure 40 psi.

Suddenly, in part (b), someone opens the valve. Pressure is no longer on the valve. If there had been a plug under the valve, at the end of the faucet, the water would immediately have encountered the plug, and the 40 psi would be applied to the plug. But there is no plug, so the water starts to flow into the vat.

As you can see from parts (b), (c), and (d), it takes time for the water to flow into the vat. By the time we get to part (d), the vat is full. The pressure on the vat in the part (d), when it is full, is 40 psi.

A capacitor behaves a lot like this vat. Instead of water flowing, current flows in a capacitor. Instead of water pressure building up, charge builds up across a capacitor, creating electrical "pressure." In both systems, flow continues until they are "full."

Look at the following figure, to see a picture of a capacitor:

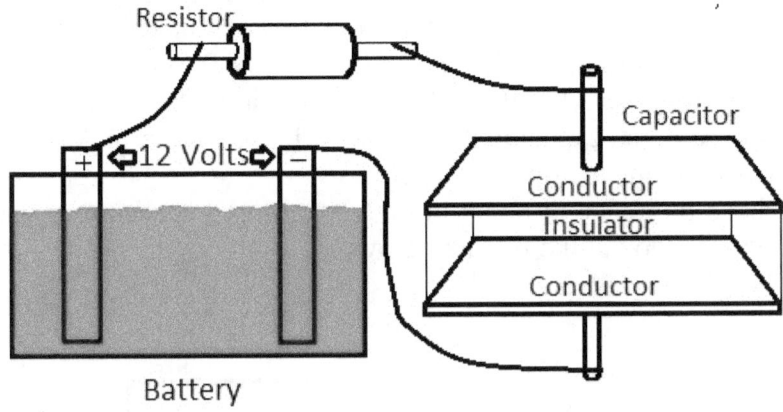

FIGURE 13-2

The capacitor is on the right, in Figure 13-2. It is ridiculously simple, and yet ridiculously complex, at the same time. It is mathematically complex. The DC circuit equation for it is:

$$V_C = V_{in} \times (1 - e^{-t/(R \times C)})$$

where e is the exponential function. Instead of the simple V=IxR equation that describes a resistor, the capacitor requires an equation based on calculus. The AC circuit equations for it use the complex number system.

And yet, the capacitor itself is so simple. As seen in Figure 13-2, it consists of two metal plates (conductors), separated by an insulator.

(The three parts of a capacitor are often formed into an inner conductive cylinder and an outer conductive cylinder, with an insulating cylinder between them.) Remember that a conductor is like a wire: current flows so easily that it offers no resistance to current flow. An insulator is the opposite: it prevents current flow; the insulator offers infinite resistance to current flow.

A capacitor is an enigma. If you stare at the circuit above long enough, you eventually realize that it is an open circuit! In a way, it's not a "circuit" at all, because current cannot make a complete "circle." It can't get through the capacitor, because the capacitor has an insulator in the middle. The capacitor looks more like an open switch, and an open switch is something we use to *stop* current flow. In fact, current never flows *through* a capacitor; it only flows *up to* its plates. Current flows through other electronic devices, but not through a capacitor. And yet, capacitors can be used like resistors in AC (voltage sine wave) circuits to control current flow.

Perhaps the place to begin in understanding a capacitor is with an open switch. Look at the picture below:

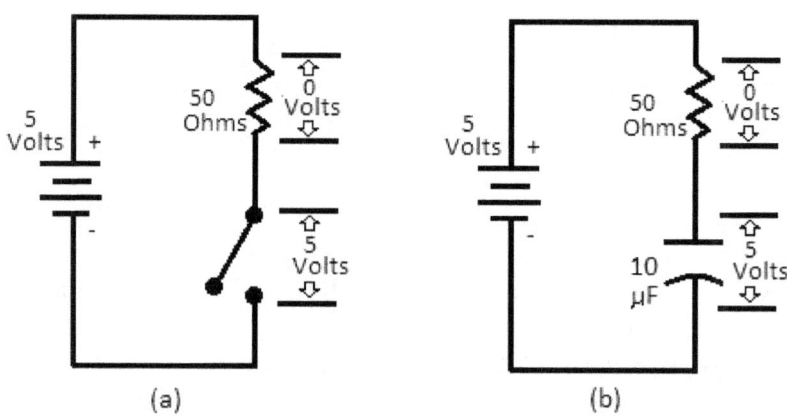

FIGURE 13-3

Hopefully, you remember the capacitor symbol from Figure 12-2, where you first saw it. It's in the lower right hand of part (b). It looks like a real capacitor: two metal plates separated by an insulator.

Just like the open switch circuit in part (a), the capacitor circuit in part (b) arrives at the same end result. Both the open switch and the capacitor wind up with 5 volts across them: the same 5 volts as the battery has. Both resistors wind up with 0 volts across them, because, in the end, no current flows through either one. No current means no work being done moving electrons through the resistor, which means no voltage, since voltage is work per coulomb of charge.

So, what is the difference between the switch and the capacitor? The difference is size. The size of the switch connections to the circuit is small. The size of the capacitor's connection to the circuit is relatively huge: it is the big metal plates that make up the body of the capacitor.

Now, we are getting somewhere. Look at the picture below:

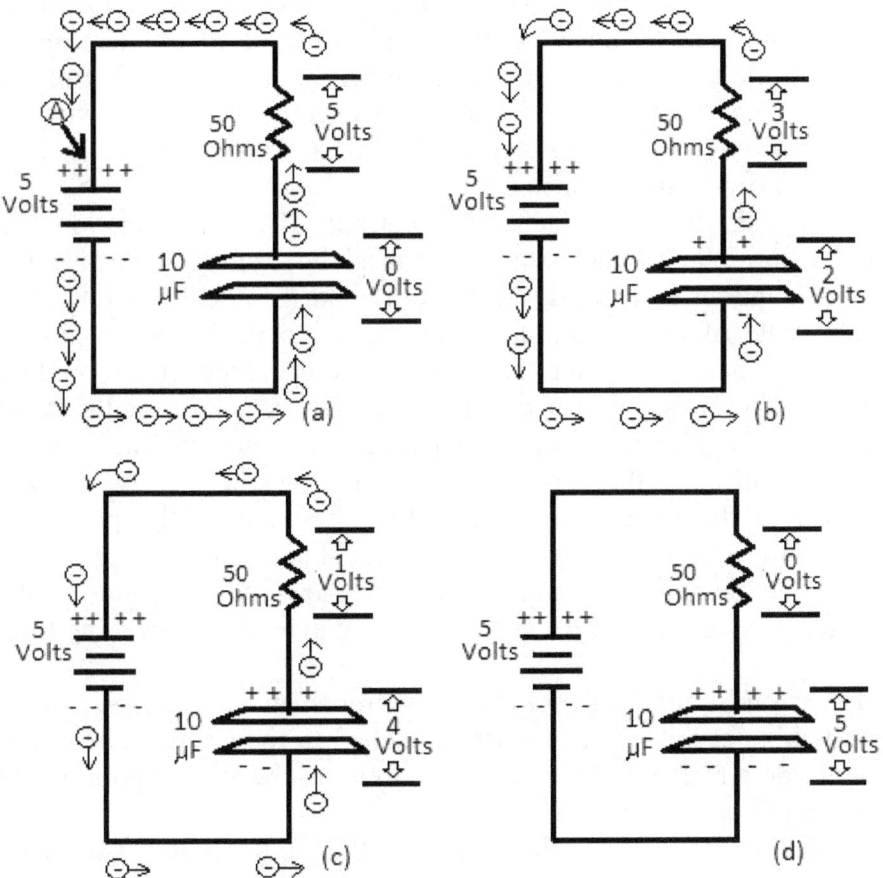

FIGURE 13-4

Figure 13-4 is the capacitor version of Figure 13-1, the water valve. In Figure 13-4, the battery and the resistor are drawn with circuit diagrams, while the capacitor is drawn as a physical capacitor. Part (a) shows the instant of time when the battery has just been connected to the rest of the circuit, by attaching a wire at point A (I could, instead, have drawn a switch being closed). Before this point in time, charge was built up on both battery plates, trying to get away from like charges. Electrons on the negative plates want to get away from other electrons. Positively charged atoms on the other plates want to get away from other positively charged atoms.

At the instant of time in part (a), the charges first get their chance. The electrons on the negative battery plate, repulsed by each other, move out into the attached wire. The capacitor, at this point in time, is neutral on each plate: all its atoms are exactly in balance with no excess positive or negative charge. So, it has 0 volts across it. But, current is flowing in this open circuit!

Current is the movement of electrons. The electrons on the negative terminal of the battery, repulsed by each other, leave the terminal, and head toward the bottom plate of the capacitor. The bottom plate of the capacitor is "attractive" to the negative charge, since zero charge is better than the negative charge on the battery terminal. The electrons in the wire head that way, too, for the same reasons, and to get away from the negative charge heading their way.

Meanwhile, on the top capacitor plate, the negatively charged electrons in the neutral atoms feel the attraction to the positively charged atoms on the positive battery plate. They start moving toward the positive plate. The electrons in the wire head that way, too, for the same reasons, and to get away from the negative charge heading their way.

Focus on Figure 13-4, part (a). Remember, this is the first instant, when the battery at point A is first attached to the circuit. The minus signs inside circles are the electrons in motion; the arrows point in the direction of current flow.

In part (a), notice the voltages around the circuit: 5 volts across the battery, 5 volts across the resistor, and 0 volts across the capacitor. We need to understand this in terms of the Law of Conservation of Energy, which we discussed in Figure 5-5.

Our electronics version of this law is called Kirchhoff's Voltage

Law. It states that the sum of all the *stored energy* in our circuit at any instant of time, must equal the sum of all the *work done* at that instant.

We know that the 5 volts across the battery is an energy input that remains fixed. It is due to the separation of charge stored on its plates, constantly replenished by the internal chemical reaction, even when charge leaves.

The capacitor is not a resistor: no work is ever done in getting current through it, because current can't go through its open circuit. In part (a), no energy is stored in the capacitor yet. So, since in part (a) the capacitor accounts for no energy or work, its voltage is 0 volts.

This implies that the voltage across the resistor must be 5 volts. This is "work done" voltage. All the energy in the battery is applied to the resistor as work to move current through it. Since I=V/R=5/50=.1, then .1 amps of current flows in this open circuit, at the initial instant depicted in part (a).

Now, things get even more interesting a short time later, as shown in part (b). Excess electrons have entered the lower capacitor plate. There are more electrons than atoms, so we have atoms that are negative ions. The lower capacitor plate takes on a negative charge, like the negative charge on a battery plate.

Also in part (b), the upper capacitor plate takes on a positive charge, like the positive charge on a battery plate. This occurs because some electrons have exited for the positive battery plate, leaving behind atoms on the upper capacitor plate that are missing an electron (positive ions).

Have you noticed that the words I have used to describe the capacitor sound like the words I have used to describe the battery? It's true. The capacitor is becoming like the battery. Recall our definition of a battery's voltage as, "a separation of charge." A separation of charge is, indeed, building up on the capacitor's plates. The difference is, the source of the charge separation is external for the capacitor: it comes from the battery's current. The source of the battery's charge separation is internal: it comes from its chemical reaction.

Figure 5-5 had one battery (B) and two resistors (R1 and R2), connected in series. We can describe the voltage across the battery as V_B, the voltage across R1 as V_{R1}, and the voltage across R2 as V_{R2}. Using Kirchhoff's Voltage Law (KVL), we described that circuit mathematically as:

$V_B = V_{R1} + V_{R2}$.

We put the voltage from the energy input device, the battery, to the left of the equal sign. We put the voltage from the work done getting through the two resistors to the right of the equal sign.

How do we create a similar equation to handle the capacitor? I have already said current never is forced through a capacitor, so no work is done to it. It is absorbing charge on its plates. It is becoming like a battery. So, its voltage belongs on the left side of the equation, with the battery, an energy (charge) storage device.

Before we put it there, though, look back at Figure 13-4. If you trace through the circuit, you will find that the charge on the capacitor points in the opposite direction as the charge on the battery. The positive end of the battery is *clockwise* of the negative end. The positive end of the capacitor is *counter-clockwise* of the negative end. So, the capacitor's voltage belongs on the left side of the KVL equation, but with a negative sign, as in $V_B-V_C=V_R$, where V_C is the voltage across the capacitor, V_R is the voltage across the only resistor, and V_B is the battery voltage. Usually, though, we move the V_C to the right side of the equation using algebra, to arrive at the equation for the circuit in Figure 13-4: $V_B=V_R+V_C$.

Comparing this equation to part (b) of Figure 13-4, we see that the voltage distribution does, indeed, obey the equation: 2 volts across the capacitor (V_C) plus 3 volts across the resistor (V_R) does in fact equal the voltage across the battery (V_B). The capacitor is acting like an anti-battery.

With this in mind, look at the number of moving electrons I have drawn in part (b) compared to part (a). I have drawn less moving electrons in part (b). In the later time depicted by part (b), less current is flowing than in the initial time depicted by part (a).

Why is this true? Because the amount of current flowing in this circuit is controlled by the resistor. Ohm's Law states that $I=V_R/R$. But, V_R has changed since part (a). V_R was 5 volts at the time of part (a), but now V_R is only 3 volts at the later time of part (b). So, the current now is only 3/50=.06 amps. Before, in part (a), it was .1 amps. The rising voltage across the capacitor subtracts from the fixed voltage across the battery, leaving less net energy (3 volts) to drive current through the resistor. The current flow rate is slowly being choked off.

We call the filling up with charge, and the rise of voltage across the capacitor, "charging the capacitor." In this circuit, the capacitor "charges" more quickly, at first, then more slowly, as time goes by. It's as if the capacitor is a bucket, and water (current) is flowing into it. Carrying this analogy farther, it's like someone is holding the hose, and gradually putting more and more of a kink in the hose, until it eventually chokes off all water flow, completely.

In part (c) of Figure 13-4, the process continues. The capacitor has charged up to 4 volts, the battery is fixed at 5 volts, so only 5-4=1 volt net energy is left to appear as the work voltage across the resistor. This produces even less current flow though R. Consequently, I have drawn even less electrons moving in part (c).

This process ends when the capacitor is fully charged. When does that happen? What does "fully charged" even mean? Well, the process ends in part (d) of Figure 13-4. The capacitor is fully charged when no more current is going in or out of its plates. Notice that I have drawn no moving electrons in part (d). The end of current flow happens when the voltage across the capacitor exactly equals the battery voltage. Since the capacitor voltage as an energy source subtracts from the battery voltage as an energy source, 5-5=0, and zero energy is available to do the work of pushing current through the resistor. Hence, there is 0 volts across the resistor. Since $I=V_R/R=0/50=0$, there is 0 amps of current flowing. The capacitor is fully charged or as full of stored current as it can get, in this circuit.

Hopefully, you noticed that I drew more and more charge symbols on either side of the capacitor as we moved from part (a) through part (d). Of course, this represents the capacitor filling up with charges, positive and negative, as it becomes more and more like the battery.

With these plus and minus symbols on the capacitor, and the capacitor fully charged to 5 volts, perhaps now is a good time to consider what it means for the capacitor be at 5 volts? How is it different from the capacitor being at 2 volts? This brings us to the essence of the concept of voltage.

Voltage is directly related to charge density. So, on both the battery and the capacitor, a higher voltage correlates to more charge being crammed closer together. The closer together (density) charge can be crammed, the stronger the force they exert on trying to get apart, on producing current flow. So, when the capacitor reaches 5 volts, it has reached the same charge density as the battery.

Can we make that the charge density that we describe as 5 volts harder to reach? Stated differently, can we do anything to increase the time it takes to get the charged packed together this closely?

Think back to the analogy of water flowing through a hose, into a bucket. What are two ways to increase the length of time it takes to fill up the bucket? One way is to make the water flow more slowly. We can do that by increasing the value of R, the resistor in the circuit. Current will flow more slowly. It will take a longer time until the capacitor reaches fully charged at 5 volts.

The second way is to get a bigger bucket. All other things being equal, a bigger bucket takes longer to fill than a smaller bucket. The capacitor is like the bucket. A capacitor can be made bigger by increasing the area of the capacitor's plates. So, if we get a bigger capacitor, there will be more room for the charges to keep their distance from each other. It will take more time, and more current will have to flow, until the charges are packed together closely enough to be called 5 volts.

These concepts are embedded in the strange equation we referred to above: $V_C=V_{in}x(1 - e^{-t/(RxC)})$. The e symbol is a function that describes how the capacitor charges up quickly at first, then more slowly as time goes by. The RxC describes how both R and C, the resistor (R) in ohms or the capacitance (C) in farads, control the charging rate.

How long does it take for the capacitor to be fully charged? From the above equation, we know that the capacitor reaches 63% of fully charged, or 3.15 volts, in R times C seconds. In our circuit, R times C is 50x.000010, or .0005 seconds. But remember, capacitors charge quickly at first, and then the charge rate gradually slows down. It takes a much longer time to finish the last 37% or 1.85 volts of charging. In fact, it takes 5 times R times C for the capacitor to go from no charge to fully charged. That calculates as 5x.0005, or .0025 seconds. If you want the 10 microfarad capacitor to take a whole second to reach fully charged, just replace the 50 ohm resistor with a 20,000 ohm one.

You can actually disconnect the charged capacitor from the circuit and walk around with it. It will hold the charge for a good while, until the electrons slowly sneak through the insulator. Insulators are not perfect, not infinite ohms of resistance.

The above experiment of carrying around a charged capacitor is not a good idea. Capacitors can hold enough energy to kill you. That's why computer power supplies sometimes have labels on them warning people not to open them, even when disconnected from 120 volt line voltage. People don't know that capacitors can store voltage even when the external input voltage is off.

You can discharge a capacitor. If you put a resistor across a charged capacitor, it will behave like a battery with a resistor across it. The electrons built up on the negative plate will scurry through the resistor, over to the positive capacitor plate. This happens because the electrons on the negative plate push away from each other, and are pulled toward the oppositely charged positive ions on the positive plate.

There's one big difference between the capacitor and the battery, though. The battery has its internal chemical reaction to replenish the separation of charge on its plates. The capacitor does not. When each atom on the capacitor's plates goes from being an ion to being a neutral atom, it is done. No internal chemical reaction restores the capacitor's neutral atoms to an ionic state.

The whole capacitor returns to neutral, to 0 volts. How long does this complete discharge take? Use the same calculation: 5 times R times C. If you want to discharge the capacitor quickly, just put a wire across it. (Remember the earlier warning about touching charged capacitors, though.) Since a wire's resistance is 0 ohms--and 5 times R times C is 0xC, or 0 seconds--the capacitor will discharge instantly, probably with a spark.

The capacitor's discharge rate through a resistor is an exponential function, just like the charge rate. This means that the capacitor discharges more quickly at first, then more slowly, until it is completely empty of charge. The equation to describe all this is:

$$V_C = V_T \times e^{-t/(R \times C)}$$

As I mentioned earlier, capacitors are very important in communications electronics, because sound, video, and radio frequency all have to do with sine waves. Our AC voltage that we access whenever we plug into an electrical wall outlet is also a sine wave.

Capacitors work strangely well with sine waves. A sine wave of voltage causes current to flow from one of the capacitor's plates to the other, in one direction. But then, the sine wave of voltage switches polarity, sucks the charge out of the capacitor, and causes current to flow from plate to plate in the opposite direction.

This forward and reverse direction of current flow in a capacitor is like the forward and reverse direction of current flow of a resistor excited by AC voltage (sine waves). From the point of view of the wires and the other components in the circuit, they have no idea that the current flowing into and out of the capacitor branch is not going *through* the branch. The capacitor is an open circuit, so the current is just going back and forth, into the plates, then back out. The only difference is a phase shift. The advantage of the capacitor over the resistor is: the capacitor consumes no power. It doesn't get hot.

One other strange thing about a capacitor is that we can increase its capacitance value--thus increasing its charging time--by changing its insulator material. The opposite charge on the two plates create an electrical field of attraction from one plate to the other. It's like two magnets with the north pole of one pulling toward the south pole of the other. In the capacitor, the insulator (called the *dielectric*) is in between; a solid insulator stops the two plates from pulling together until they touch.

The electrical force field produces a strange effect on the insulator. It causes the individual molecules in the insulator to turn, and even stretch. The insulator's molecules become more positive on the side closer to the negative plate, and more negative on the side closer to the positive plate. The molecules, in turn, create a counter-field produced inside the insulator. The positive charge in the insulator's turned and stretched molecules creates a positive electrical field that reaches into the negative plate and cancels out some of its negative strength. The negative charge in the insulator creates a negative electrical field that reaches into the positive plate and cancels out some of its positive strength. With each plate's repulsive energy weakened, more charge is needed for it to reach 5 volts (in our example). A greater charge density is needed to arrive at a net mutual charge repulsion which we call 5 volts.

Some insulator materials are simply better than others at creating the above effect. The molecules in some insulators stretch, turn, and align better than molecules in other insulators. With a "better" insulator, current has to flow longer, to fill the capacitor.

Less subtly, the capacitor's plates themselves cause this same effect on each other. The positive plate's positive electric field subtracts from the negative plate's negative electric field, and vice versa, allowing more current to flow before 5 volts is reached. Furthermore, positive and negative electric forces of attraction and repulsion increase with less distance between charges (an inverse square law). So, bringing the positive and negative plates closer intensifies this effect, increasing capacitance and charging time.

CHAPTER FOURTEEN
THE CRYSTAL CLOCK

While we are looking at strange effects, let's look at something even more strange: the piezo-electric crystal. The piezo-electric crystal is the strangest component you never heard of.

First of all, the reason we will discuss the piezo-electric crystal is that it is used to make a better system clock. The clock in Figure 12-2 is sufficient in some microprocessor designs. It is very inexpensive. But it has some shortcomings that make it undesirable in some applications.

First, heat affects the circuit in Figure 12-2. So, as temperatures change around the components, their values change, which changes the oscillatory frequency. Also, designing the circuit to a specific frequency is difficult, because the accuracy of capacitor values is not very good. A 10 microfarad capacitor might be plus or minus 20% of its rated capacitance value.

So, we often use clocks designed around piezo-electric crystals: crystal clocks, as they are called. Though not perfect, they tend to be less temperature-sensitive and more accurate than clocks like the ones in the circuit of Figure 12-2.

Here's where the weirdness starts. The piezo-electric effect is the creation of voltage by pressing on a crystal. Quartz crystals work well for this. We know that voltage is a separation of charge. Well, if you cut a crystal just right, and apply a pressure to it, charge gets separated, and a small voltage is produced across the crystal.

But, the weirdness does not end here. The opposite effect also exists. That is, if you apply a voltage across a crystal, it will compress.

So, what we have is two opposite weirdnesses, which we combine together to create an oscillator:

1.) Voltage makes compression
2.) Compression makes voltage

Here's how it works. A voltage source is applied to opposite sides of the crystal. Let's use the example of a battery connected, through wires to the crystal.

Now that you understand about capacitors, think of the crystal as if it were a capacitor. A positive charge builds up on the side of the crystal connected to the positive battery plate. A negative charge builds up on the side of the crystal connected to the negative battery plate. As with the capacitor, an electric field of attraction builds up from the positive to the negative side if the crystal. They try to move toward each other. Of course, the center part of the crystal is in the way. So, the center part of the crystal gets squeezed by the outer parts of the crystal.

This explains how step 1 works: how voltage makes compression. We don't squeeze the crystal with a vice. We squeeze it with an electric field, which tries to pull the sides of the crystal towards each other.

Step 2 says that this compression causes a separation of charge in the crystal. Negative charge in the crystal moves toward the positive side of the crystal, and positive charge in the crystal moves toward the negative side of the crystal. This charge from within the crystal cancels out the charge from outside the crystal supplied by the battery. To summarize, the battery puts charge on the sides of the crystal; the crystal compresses; the compression creates charge that cancels the charge on the sides of the crystal.

Of course, with the charge gone on the sides of the crystal, the electric field of attraction is also gone between the two sides of the crystal. They stop squeezing the insides of the crystal. The crystal relaxes to its natural, unsqueezed state. The separation of charge that resulted from the squeezing is gone, too.

In fact, everything has gone away. Everything has returned to where we started. But, the battery is still there, and the battery is still connected to the crystal. So, the entire process starts all over. Steps 1 and 2 start all over. The battery once again causes a separation of charge on the sides of the crystal, etc. The picture below summarizes the explanation:

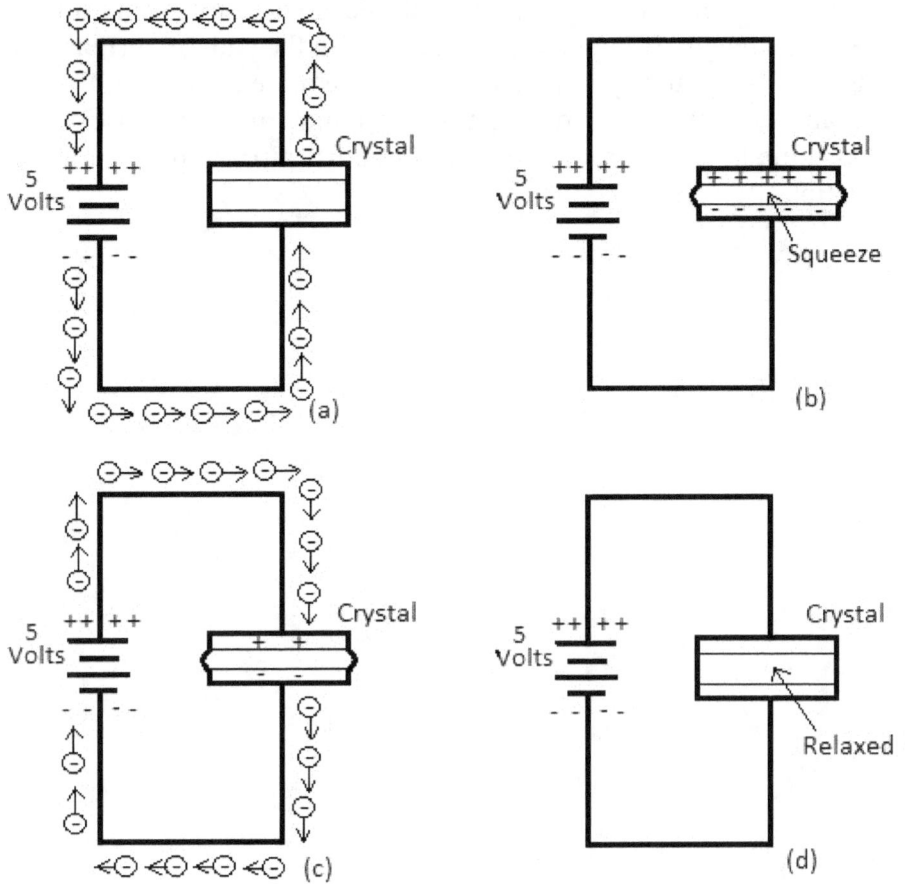

FIGURE 14-1

Part (a) is the voltage applied. Part (b) is the electric field from the charge separation causing the compression of the crystal. Part (c) is the compression creating a counter-voltage. Part (d) is the decompression and discharge of the crystal, ready to restart the cycle.

The picture lets you see the oscillation. In Figure 14-1 part (a), current flows counter-clockwise; in part (c), it flows clockwise. In part (b), the crystal has maximum positive charge on top on top and negative charge on the bottom; in part (c), the crystal's counter-voltage is neutralizing the charge on the plates.

A clock built with a quartz crystal needs some extra circuitry to support this magic. If you imagine a resistor in the path of current flow in Figure 14-1, you may realize that a voltage more akin to an AC sine wave appears across it. So, additional circuitry turns the oscillation into a digital, or square wave. The circuitry also helps sustain the oscillation.

CHAPTER FIFTEEN
THE FLIP-FLOP

Let's return to our microprocessor system of Figure 8-1. Remember, we just addressed the question of how a microprocessor does things sequentially. We showed that a microprocessor uses a digital clock, to break steps down into ticks of a square wave cycle. But, how do we convert this string of voltage changes on one wire, the clock output, into separate voltages that enable--in order--fetch, then decode, then execute?

To answer that question, first look back at figures 10-1 through 10-3. They depict our 2-to-4 binary decoder. What if we had a device that could count our clock pulses, and output that count value in binary? We could feed that binary output as an input to our 2-to-4 binary decoder. We would get sequential pulses out of our 2-to-four binary decoder: Q0, then Q1, then Q2, then Q3, then back to Q0, etc.

We'll explore that design later, but the name for the device we need is the *binary counter*. We have another need for a binary counter. In our review of our microprocessor system in Figure 8-1, we referred to the program counter. This was a binary counter, too. It looks like an understanding of a binary counter is an important part of understanding how a microprocessor works.

To create a binary counter, we must first introduce a re-design of our D latch. Our counter will require connecting the NOT Q output back into the D input when E is enabled. That trick just won't work with a D latch. If you review the D latch circuit way back in Figure 7-10, you will find the "dog chasing its tail" effect, again. A low NOT Q fed back to D immediately makes NOT Q high; which, feeding back to D immediately makes NOT Q low, etcetera, ad infinitum.

We must turn our D latch into a D flip-flop. What's the difference? In a latch, as long as the E (enable) input is high, the Q output replicates the D input. If the D input changes a hundred times, the Q output changes a hundred times. Whatever value was on the D input and Q output when the E window closes (goes low) is the value stored at the Q output.

For the flip-flop, we get rid of the E (enable) input and replace it with the clock, or CLK, input. Sounds like it was designed specifically as the destination for the digital clock we just studied, doesn't it? Next, we decide if it will be positive edge triggered or negative edge triggered. We choose positive edge triggered.

Our new, positive edge triggered flip-flop only allows the value at the D input to transfer and be stored at the Q output at one instant of time. That instant is: every time the voltage at CLK rises from 0 volts to 5 volts. No transfer or storage happens while CLK sits high; a latch does that, and this is not a latch. No transfer or storage happens while CLK sits low. No transfer or storage happens when CLK falls from 5 volts to 0 volts.

It's called, "positive edge triggered" because a positive-going CLK voltage initiates the change. Neither a negative nor a positive voltage initiates the change. A voltage leaving negative and going toward positive initiates a transfer and storage of the D input value into the Q output.

The picture below will help clarify all this, if it is still unclear:

D Flip-Flop

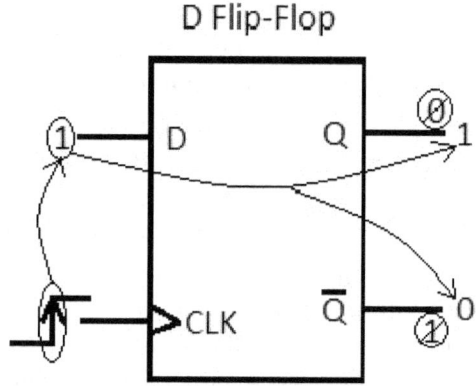

FIGURE 15-1

Figure 15-1 shows the logic symbol for the D flip-flop. Notice the CLK label that replaces the E symbol we used for the latch. Also, notice the little wedge next to the CLK label. This wedge lets us know that this device is edge-triggered. If there was a little bubble to the left of the wedge, that would tell us this device is negative edge triggered. This one is positive edge triggered.

Look carefully at what is coming into the CLK input. Read the input as voltage, vertically; with 0 volts at the bottom, and 5 volts at the top. And, read it as time, horizontally; with past to future from left to right, respectively. Thus, early in time, the voltage is 0 volts. A transition from low to high happens, along the up-arrow. Later in time, after the transition, the voltage is 5 volts.

Notice how I have the up-arrow circled. This is the key instant, the exact and only time the D flip-flop is opened up for storage. Follow the curving arrows. They indicate that, at the critical time of the positive-going voltage edge on CLK, the 1 on D transfers to the Q output. As always, the NOT Q output becomes the opposite logic level of Q, so it becomes 0. The values crossed out are the old values of Q and NOT Q, from before the rising edge triggered their changes.

The picture below shows how to put NAND gates together to create a D flip-flop:

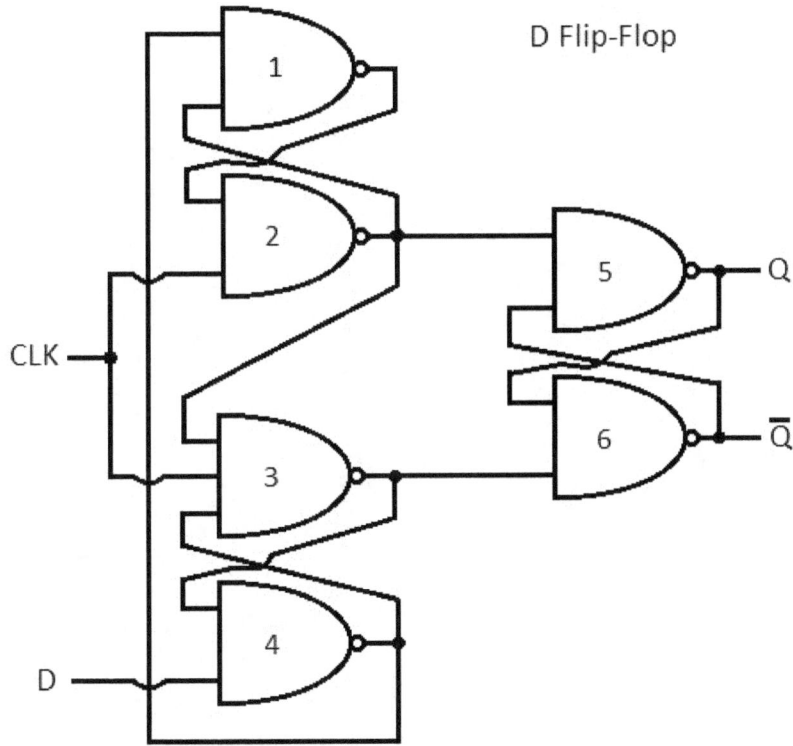

FIGURE 15-2

I won't go through, step-by-step, how the D flip-flop triggers only on the rising edge of the CLK input. It would take too long, and you can do it for yourself, if you like. Notice a few things, though. The cross-coupled NAND pairs (1 and 2, 3 and 4, and 5 and 6) are RS latches. This time, they are made from NAND gates instead of NOR gates. They function the same way, except the NAND pairs have the S inputs across from the Q output, whereas the NOR pairs have the R inputs across from the Q output.

With this in mind, Figure 15-2 almost simplifies to three interconnected RS latches. I say "almost," because NAND gate 3 is a little different. It is a 3-input NAND gate instead of a 2-input NAND.

So, I said we needed this D flip-flop so that we can connect the NOT Q output back to the D input. It will become an essential part of our binary counter, and it overcomes the, "dog-chasing its-tail," problem. Let's make this connection, and see what we get:

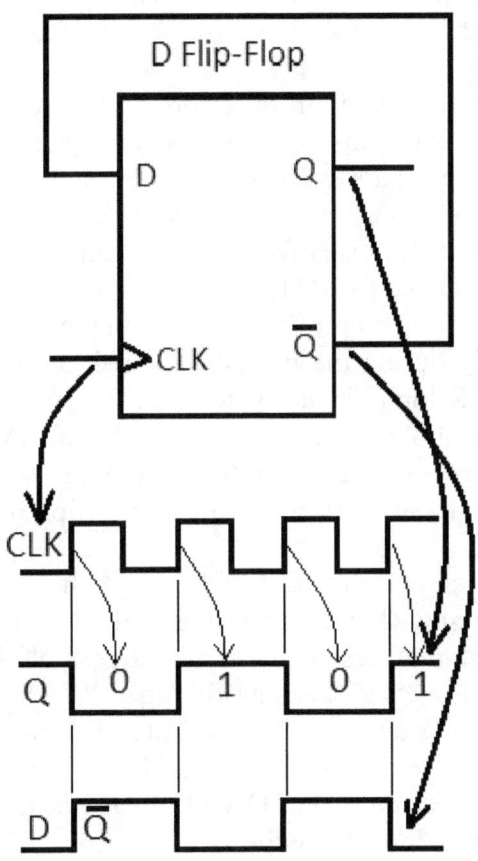

FIGURE 16-1

Let's get comfortable with this figure. At the top is our D flip flop, with the NOT Q output connected back to the D input. At the bottom are three voltage waveforms. These are referred to as, "timing diagrams." I will use this term, because it emphasizes the fact that the horizontal axis represents time. Earlier time starts on the left. Later--or future--time moves to the right. Also, the three waveforms align vertically. You can put your finger on any one waveform, then move it up or down to see the exact voltage state of the other two waveforms at the same instant of time.

You can also call this a voltage waveform, because the vertical axis represents voltage. Since this is digital voltage, we really only have two possible voltages. The top of each square wave is 5 volts (logical 1, or high), and the bottom of each square wave is 0 volts (logical 0, or low).

The thick arrows show from where each waveform originates. Remember, the waveforms are vertically aligned so that we can observe when events happen in relation to each other, at the same instant of time. We want to see the effect of tying the NOT Q output back to the D input. Because of this connection, the bottom waveform is labelled both as D and as NOT Q.

We know that a flip-flop transfers the voltage on the D input to the Q output, but only during the instant of transition, the rising edge, of voltage on the CLK input from low to high. But what is on the D input is the NOT Q output. And, the NOT Q output is always *opposite* of the Q output. So, every rising edge on the CLK input forces the Q output to its *opposite* logic state. We call this state-flipping on every clock edge, "toggling."

The timing diagram shows this. Look at the line labelled CLK. Then look at the rising (low to high) edges of CLK. I have placed thin, vertical alignment markers under these edges. Notice how Q only changes state during these instants of time. Watch how the state of *NOT Q* before this instant becomes the state of *Q* after this instant. See how when Q goes to one logic state, NOT Q always goes to the opposite state. This ensures that Q will always change to its opposite logic state, at each rising CLK edge.

What's the point of all this? Follow the skinny arrows, leading from each rising edge of CLK. Reading left to right, the skinny arrows are telling us that the first edge makes the Q output 0, the second edge makes the Q output 1, the third edge makes the Q output 0, etc.

The flip flop is counting! It's only a 1-bit counter, like the least significant bit in a larger counter. But remember, available binary numbers end at 1. The next number, after 1, is: reset to 0, then carry a 1 to the next more significant bit. We have just accomplished the "count up to 1" part, and the "reset to 0" part. All we need to do now is accomplish the "carry to the next more significant bit" task.

So let's do that. See the next picture:

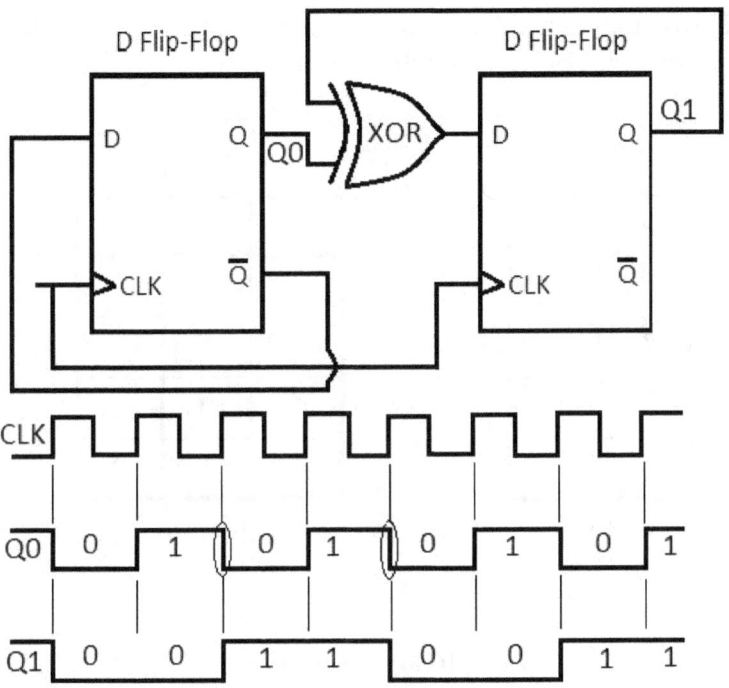

FIGURE 16-2

Now we have two Q outputs, so I have labelled them Q1 and Q0. A 2-bit binary counter counts like this: 00, 01, 10, 11. Then it goes back to 00, and counts the same sequence. These binary numbers correspond to decimal numbers 0, 1, 2, 3. Q1 is our most significant bit, and Q0 is our least significant bit.

It is counting! Prove it to yourself. Look at the Q1 and Q0 lines of the timing diagram. Read from left to right, and read Q1 before Q0. You will see that the counts are: 00, 01, 10, 11, 00, 01, 10, 11.

To accomplish this design, we added a second D flip-flop, on the right. Its output is Q1, our most significant bit. We connected its Q1 output along with the unchanged output of our original flip-flop (re-labelled Q0) as inputs to an exclusive-OR gate. We then connected the output of the exclusive-OR gate to the input of the new flip-flop.

We know the circuit counts properly, but how does it work? Perhaps the picture below will help:

Counter				XOR Gate		
Q1	**Q0**			**A**	**B**	**Y**
0	0			0	0	0
0	1			0	1	1
1	0			1	0	1
1	1			1	1	0
(a)				(b)		

FIGURE 16-3

Part (b) of Figure 16-3 is the truth table for an exclusive-OR gate. We will need to refer to it, since the Q0 and Q1 outputs both enter it. Part (a) shows the consecutive counts of the counter, out of Q1 and Q0.

In part (a), we know that the Q0 column alternates in a 0-1-0-1...etc. pattern. We called it toggling, and learned about it in Figure 16-1. What we must do is explain the new part, the Q1 column. Look at the first row of *part (a)*. It is 0-0, the output of Q1-Q0. 0-0 is fed into the exclusive-OR gate (XOR). The first row of *part (b)* tells us that the output of the XOR--thus the D input to the Q1 flip-flop--is 0. This means that, upon the next rising clock edge, a 0 will come out of Q1.

That's why in part (a) I have a big circle around 0-0 in row 1, with an arrow pointing to a square in row 2 with a 0 inside it. The 0-0 (per the XOR truth table) makes Q1 equal 0 in the next row, on the next clock pulse. In row 2 of part (a), the 0-1 (per the XOR truth table) makes Q1 equal 1 in row 3, on the next clock pulse. In row 3 of part (a), the 1-0 (per the XOR truth table) makes Q1 equal 1 in row 4, on the next clock pulse. In row 4 of part (a), the 1-1 (per the XOR truth table) makes Q1 equal 0 in row 1, on the next clock pulse.

Now that you get the idea of what counters are and how they work, I'll skip the detailed explanations of the next ones. A microprocessor's program counter can be pretty big. For a big addressable memory space, a 64-bit counter is not uncommon. I'll show you a circuit diagram for a 4-bit counter, below:

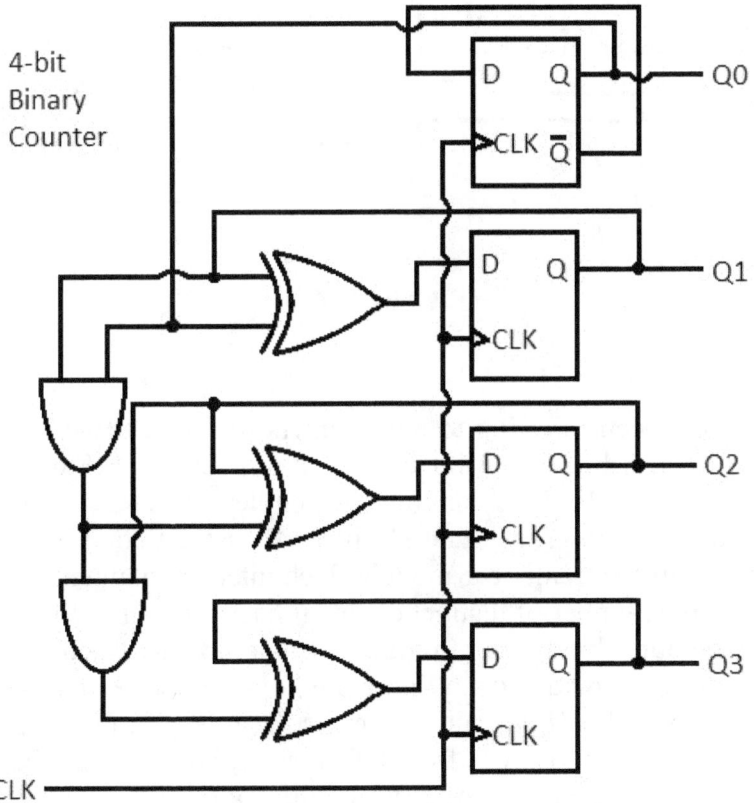

FIGURE 16-4

The above counter repeatedly counts from 0000 to 1111, so it has 16 distinct counts: 2^4. The top two flip flops, for Q0 and Q1, are the same as in Figure 16-2. To expand this counter to more than 4 bits, just follow the same pattern. Add more AND's, XOR's, and D flip-flops with the same wiring pattern as in Figure 16-4.

For our fetch-decode-execute sequence, we need a counter that counts like this: 00-01-10. For that, we can use the following circuit:

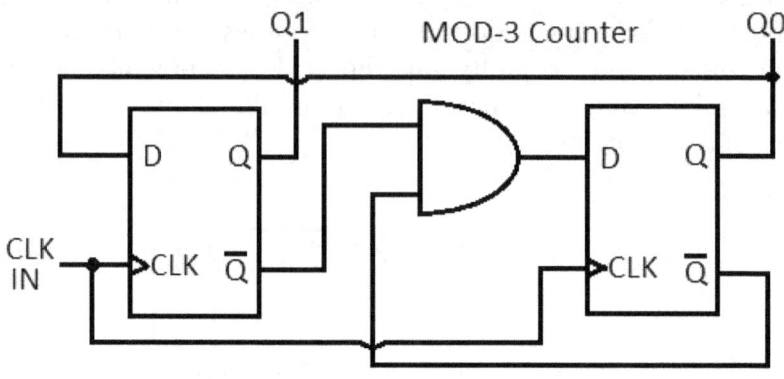

FIGURE 16-5

We have been referring to our counters in figures 16-1, 16-2, and 16-4 as 1-bit, 2-bit, and 4-bit counters, respectively. If we call the counter in Figure 16-5 a 2-bit counter, people will tend to think it has $2^2=4$ distinct counts. But it doesn't. It has 3 distinct counts: 00-01-10. We should instead call it a modulo-3 counter. A counter's modulus number is the number of distinct counts it has. We could, for example, have called our 4-bit counter in Figure 16-4 a modulo-16 counter.

Be careful if you decide to figure out how the above mod-3 counter works. Its weighted bit-values are backwards from our previous counters. The circuit flow is from left to right, but the least significant bit, Q0, is on the right, and the most significant bit, Q1, is on the left.

Now we can pick up the thread we left when we decided we needed a digital counter. We asked how we could convert a string of voltage changes on one wire, a clock, into separate voltages that enable--in order--fetch, then decode, then execute. Our solution is pictured below:

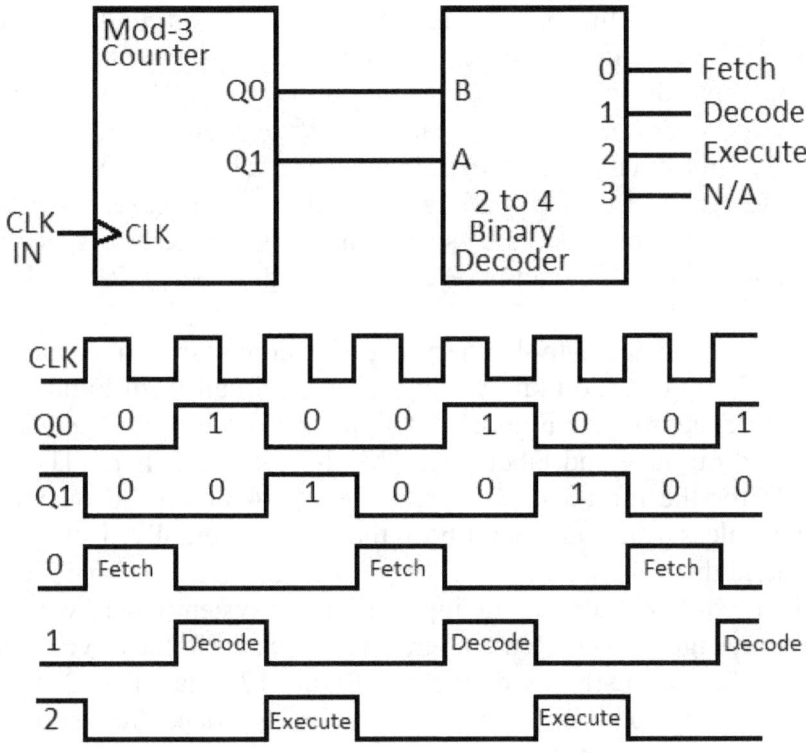

FIGURE 17-1

Let's skip directly to the bottom three lines of the timing diagram in Figure 17-1: the ones labelled 0, 1, and 2. They are the outputs of the lines out of the 2-to-4 binary decoder labelled 0, 1, and 2. These three lines are the fruits of our labor: our goal.

Make sure you understand what these three lines accomplish. Remember that the horizontal axis is time, and the vertical axis is voltage. As we move forward in time (left-to-right), we encounter the 0 line going from 0 volts to 5 volts. This line is sent to the fetch circuitry to enable its action. As we move further ahead in time, the 0 line goes to 0 volts, disabling the fetch circuitry.

Next, the 1 line goes high. This line is sent to the decode circuitry to enable its action. As we move further ahead in time, the 1 line goes low, disabling the decode circuitry.

Finally, the 2 line goes high. This line is sent to the execute circuitry to enable its action. As we move further ahead in time, the 2 line goes low, disabling the execute circuitry. Going farther ahead in time, we observe this pattern repeating itself, over and over.

The 2-to-4 binary decoder is already familiar to you. We saw it in 10-3, and we used it as our address decoder. In this case, we are not using the output marked as 3, so its output is labelled N/A, for "not applicable." (For efficiency, we could have designed a 2-to-3 binary decoder.)

The input to the 2-to-4 binary decoder comes from the output of the mod-3 counter. You know this counter's circuit from Figure 16-5. We have encapsulated Figure 16-5 into a box that only reveals the inputs and outputs, and labelled it, "Mod-3 Counter," here. The logic symbols for the mod-3 counter and the 2-to-4 binary decoder mean nothing unless you understand how the work, internally. But, you do understand them.

The mod-3 counter gets its input from our system clock, which we covered in Figure 14-1. Remember, it is a square digital wave, as seen in the top line of the timing diagram in Figure 17-1, labelled CLK.

Feeding the clock into the mod-3 counter produces its Q0 and Q1 outputs. We know that the mod-3 counter counts in the pattern, 00-01-10-00-01-10, repeatedly. You can see the counting pattern in the second and third lines of the timing diagram in Figure 17-1, labelled Q0 and Q1. I have labelled the Q0 and Q1 timing diagram lines with 0's and 1's, so that you can more easily observe them counting. Notice how things in lines Q0, Q1, 0, 1, and 2 of the timing diagram only change during the rising edge of the CLK line.

We already know all of the above things, separately. Let's try to understand how all of this works together. We have three subsystems, when you count the clock. The clock creates time increments for us, in our digital world. The mod-3 counter counts the rising edges of the clock, turning it into a 2-bit pattern of 00, then 01, then 10 (like a digital 0, 1, 2), then back to 00, to repeat the cycle.

Then, when the 00 is fed into the 2-to-4 binary counter, its 0 output goes high. When the 01 is fed into the 2-to-4 binary counter, its 1 output goes high. When the 10 is fed into the 2-to-4 binary counter, its 2 output goes high. You can see this clearly, in the above timing diagram.

The 2-to-4 binary counter takes the 2-bit counts that change with time, and converts them into single pulses that change with time. Only one pulse is high, or true, at a time. The pulses are sequential, in time. Each one turns on one action at a time: fetch, then decode, then execute.

If we like, we can encapsulate the sequencer in Figure 17-1. We would hide the mod-3 counter and 2-to-4 binary decoder inside a box, label it, "3-bit sequencer," and show only the inputs and outputs. A picture of this is shown below:

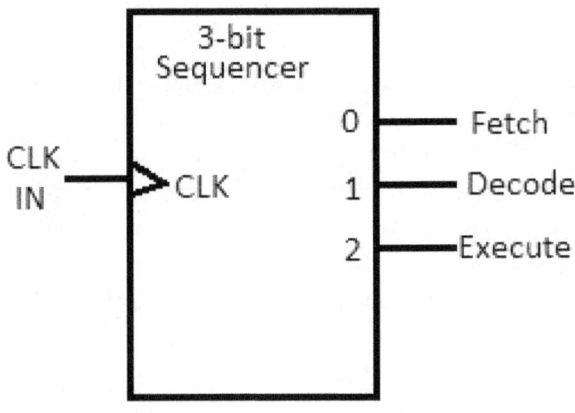

FIGURE 17-2

Let's again take a moment to appreciate how digital electronics builds complexity out of simplicity. At the highest level, we start with one box--one logic diagram--with many inputs and outputs. This box is the extremely complex microprocessor. Open that box, and you see a number of subsystem, one of which is this 3-bit sequencer, which generates the fetch-decode-execute sequence. Open the 3-bit sequencer box, and you find a mod-3 counter box and a 2-to-4 binary encoder box, connected as in Figure 17-1.

Let's follow the mod-3 counter path first. Open the mod-3 counter box, and you find two D flip-flops and an AND gate, connected as in Figure 16-5. Open a D flip-flop box, and you find six NAND gates, connected as in Figure 15-2. Open up an AND box, and you find three of our simple, fundamental building blocks, the transistor/resistor circuit, connected as in Figure 6-23. Open up a NAND box, and you find five of our simple, fundamental building blocks, the transistor/resistor circuit, connected as in Figure 6-27. To see the transistor/resistor circuit, look at Figure 6-7.

Let's follow the 2-to-4 binary encoder path next. Open the 2-to-four binary decoder box, and you find two inverters and four AND gates, connected as in Figure 10-1. Open up an AND box, and you find three of our simple, fundamental building blocks, the transistor/resistor circuit, connected as in Figure 6-23. Open up the inverter triangle, and you find our transistor/resistor circuit. To see the transistor/resistor circuit, look at Figure 6-7.

Before we move on, take one more look at Figures 17-1 and 17-2. In these circuits, we have claimed that, simply by being in the high state, any one of the outputs of the sequencer allows an event to happen. That is, a high output turns on some other circuitry to allow a fetch, or a decode, or an execute. Let's make sure you understand and appreciate this.

Actually, you already understand this. We have been "enabling" circuitry all along. We have been designing circuits that do nothing unless an enable line is pulled high.

For example, our D latch in Figure 7-10 had an enable input. Our tri-state buffer (Figure 7-20) and tri-state inverter (Figure 7-19) also had enable lines. So did our bi-directional latch in Figure 7-22. One look at our simple microprocessor-based system with its address decoder (Figure 10-4) should prove to you that one voltage pulled high can select one of several subsystems.

Or, go back to the pass/block circuit of Figure 7-11. It shows a simple AND gate, acting as an on/off switch for a clock. The clock comes out of the AND if the enable input of the AND is high; the clock does not come out if the enable input is low. Realize that, if we wanted to enable/disable any of our counters in figures 16-1, 16-2, 16-4, 16-5, 17-1, or 17-2, all we would need to do is make the input clock first pass through an AND gate, controlled by an enable voltage on one input, as in Figure 7-11.

All the above design techniques allow us to turn on or off entire subsystems with one wire--usually called "enable"--set to high for enable or low for disable. It's like a light switch, turning circuits off or on. Remember this as we move into the next section, in which multiple subsystems will be on or off.

CHAPTER EIGHTEEN
INSTRUCTION DECODING

Now, let's return to our basic microprocessor system of Figure 8-1. We shall next consider how our microprocessor decodes and executes a command. We say that digital electronics systems are wonderful. But, we claim that the digital electronics system called a microprocessor is even more amazing, because it is "*programmable*." And, it is "*re-programmable*."

In other words, a simple digital system performs one function, and acts as one machine. A microprocessor can run one program which makes it perform one function, as one machine. Then, it can run another program causing it to perform a different function, as if it were a different machine. Theoretically, it can run an infinite number of programs, so it can be an infinite number of machines. Have you used a cell phone, recently?

It does all this because it can decode and execute commands. Each command, in a sense, makes the microprocessor a slightly different machine. I will show you how it does this, shortly.

Let's start with an analogy. Think of a very big corporation, like GM. They make transportation vehicles: cars, trucks, SUV's, minivans, etc. They stock a lot of subsystems: transmissions, drive trains, suspension systems, exhaust systems, engines, wheels, gears, lights, etc.

Not every one of their vehicles is 100% unique. By this I mean that, hypothetically, the same ignition switch could be in the Avalanche, the Silverado, the Escalade, and the Sierra, but not in other GM vehicles.

So, GM can create a car called the Goofmobile by using subsystem 348 for the transmission, subsystem 59 for the drive train, subsystem 678 for the suspension, subsystem 12 for the exhaust, etc. They can then create a different car called the Daffy Truck by using subsystem 322 for the transmission, subsystem 72 for the drive train, subsystem 611 for the suspension, subsystem 18 for the exhaust, etc. The two vehicles might have some subsystems in common, like the lights, seatbelts, airbags, etc.

The point of this analogy is, once the vehicle is designed, it is done. It is one vehicle, one design. Every vehicle of the same make and model is the same. If you want a different kind of vehicle with a different design, made from different subsystems, you have to buy a different vehicle based on a different design, made from different subsystems.

Microprocessors are better than this. When the microprocessor decodes a command, it *actually interconnects different subsystems together, on the fly!* These subsystems then work together to execute the command's task.

This is truly amazing. It's as if GM could redesign and recreate a new vehicle for you every nanosecond, made from different combinations of subsystems, while you are driving it!

How does the microprocessor do this? Look at the picture below:

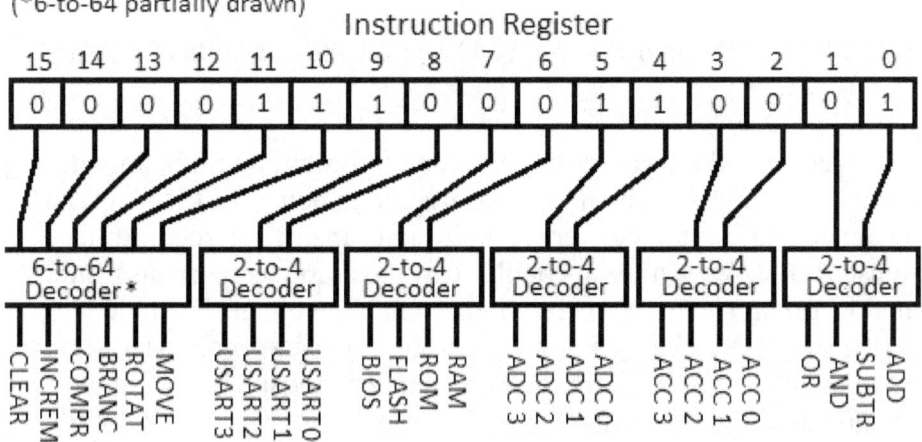

FIGURE 18-1

The top of Figure 18-1 shows the microprocessor's 10-1instruction register, with an instruction inside. The instruction is there due to the fetch process, whereby the command was copied to the instruction register from command memory. The next step is the decode process.

The command in the instruction register is 0000111000110001. Remember, the 0's are really 0 volts, and the 1's are really 5 volts. They are stored in a 16-bit latch called the instruction register.

Individual bits from the instruction register become inputs into different decoders. (You are familiar with the 2-to-4 decoder from figures 10-1 and 10-3.) Each 2-to-4 decoder has the ability to enable one of four subsystems. Not every 2-to-4 decoder's subsystem is required for every command. For example, few commands will use the 2-to-4 decoder that activates ADC 0, ADC 1, ADC 2, or ADC 3.

The 6-to-64 decoder is only partially shown. It uses 6 bits from the instruction register to choose 1 of 64 circuits to execute specific commands. I have only shown 6 of the 64 outputs: the ones that enable the MOVE, ROTATE, BRANCH, COMPARE, INCREMENT, and CLEAR circuits. Each of these specific circuits executes the command by accessing the data from the circuits enabled by the 2-to-4 decoders.

Let's say that the designer of the circuitry for this microcontroller created it so that the binary command 0000111000110001 executes as follows:

SUBTRACT the contents of ADC 3 from the contents of USART 2, storing the results in ACCUMULATOR 2.

What does this even mean? Well, the program might be monitoring the temperature of a chemical mixture that is being heated. An ADC is an analog-to-digital converter, built into many microcontrollers. A transducer--like a thermocouple--can convert temperature into an analog (many level) voltage. If that analog voltage is input into an ADC, the ADC changes it into a binary string of voltages--a binary number--that the microprocessor understands. The higher the temperature, the higher the binary number. *In short, the ADC's value is the temperature of the chemical mixture.*

The reason the program might be monitoring the temperature is that it might want to compare it to a shut-off temperature. Where will it get this shut-down value? Well, for total flexibility, the user of this microprocessor should be able to choose the shut-off temperature at any time. The user could enter it into a keyboard that attaches to one of the serial input ports. The kind of integrated circuit (IC) that converts a serial (one bit at a time) binary input into a parallel binary output (think of an 8-bit latch) is called a USART (universal synchronous/ asynchronous receiver/transmitter). Many microcontrollers have built-in USARTs. *In short, the USART's value is the shut-off temperature.*

Of course, when the temperature of the chemical mixture reaches the shut-off temperature, the microprocessor will need to turn off the heat source to the chemical reaction. How does the microprocessor know that the temperature of the chemical mixture has reached the shut-off temperature?

One way is to subtract the temperature of the chemical mixture from the shut-off temperature. If the answer is zero, the chemical mixture is at the shut-off temperature, so it's time to shut off the heat source. In terms of the devices available to the instruction register through the 2-to-4 decoders, it's time to *shut off the heat source when subtracting the contents of ADC 3 from USART 2 equals zero.*

Let's go through this command, bit by bit:

Bits 1 and 0 of the instruction register are set to 0 and 1, respectively, thereby enabling the subtraction circuitry.

Bits 3 and 2 of the instruction register are set to 0 and 0, respectively, thereby enabling the accumulator 0 circuitry.

Bits 5 and 4 of the instruction register are set to 1 and 1, respectively, thereby enabling the analog-to-digital converter 3 circuitry.

Bits 7 and 6 of the instruction register are set to 0 and 0, respectively; however, this instruction does not use any memory circuitry.

Bits 9 and 8 of the instruction register are set to 1 and 0, respectively, thereby enabling the universal synchronous/asynchronous receiver/transmitter 2 circuitry.

Bits 15, 14, 13, 12, 11, and 10 of the instruction register are set to 0, 0, 0, 0, 1, and 1, respectively, thereby enabling the compare circuitry.

This is a fine example of the microprocessor building a "car" (circuit) on the fly. Because of which bits are in the instruction register, USART 2 and ADC 3 are enabled, their contents subtracted in the enabled subtraction circuitry, and the answer stored in the enabled accumulator 2. Out of 84 subassemblies (circuits) available to the microprocessor, the instruction register has assembled 5 of them to make a subtraction machine. A different command builds a different "car" (circuit).

In case you are curious, here's what can happen next (depending on the microprocessor design) to wrap up the task of leaving the heater on or turning it off. (It's an advanced topic, but I put it here for a sense of completeness, and for the curious.) If you look closely, you will see that this command does a little more than just subtraction. It also activates the compare circuitry. If the value in the accumulator 2 is all zeroes, the compare circuitry sets a 1-bit flip-flop called "zero" to 1. If the value in the accumulator 2 is not all zeroes, the compare circuitry sets the "zero" flip-flop to 0.

The next command can be a conditional *skip* command. This command tests the "zero" flip-flop. If the "zero" flip-flop contains a 0, the skip circuitry adds an extra count to the program counter. Otherwise, if the "zero" flip-flop contains a 1, the skip circuitry does not add an extra count to the program counter.

This little trick makes the microprocessor skip the fetching of the next command from program memory if the chemical mixture is not up to the shut-off temperature. Normally, the microprocessor fetches the next command from *program counter + 1.* (Remember, the program counter automatically increments by one after every fetch.) The skip circuitry makes the next command fetch come from *program counter + 2,* if the value in accumulator 2 is not all zeroes.

However, if the value in accumulator 2 is all zeroes, that means the chemical solution has reached the shut-off temperature. The skip circuitry does not add an *extra* count of +1 to the program counter. The next command, from *program counter + 1,* gets fetched. This command would load a totally new value into the program counter, redirecting the microprocessor to fetch the next command from an area of code that turns off the heater.

What is in *program counter + 2*? It probably contains code to loop back around to test the temperature again, waiting for the time to shut off the heater.

Computer programmers may recognize that this as the basis for the way computers actually accomplish the FOR loop. An accumulator register can be loaded with number, decremented (subtract one), and tested as above to see if it has reached zero.

You will notice my frequent use of the word, "might," as in, "here's what the compare circuitry might do." That's because I am making up microprocessor circuits and programs as I go along. I am tailoring them to be good designs for teaching you how microprocessors work. There are many ways to design a microprocessor, but they all have a similar basic design.

Thus, I have tried to make a generic microprocessor for you to learn. Remember, our goal is for you to understand how microprocessors work, not to become a microprocessor designer. In fact, there are not that many jobs for designers of microprocessors.

Most computer engineers design systems based upon off-the-shelf microprocessors: microprocessors already designed and sold by the millions or billions. The computer engineer has a general knowledge of how microprocessors work, then applies that knowledge to learn the specifics of how that particular off-the-shelf microprocessor works, then designs a product based on that design.

That being said, let's look at another way that a microprocessor designer "might" create circuitry that decodes the contents of the instruction register. Let's assume (to fit the design on one page) that the command is only 8 bits wide, so the instruction register is 8 bits wide. Like in Figure 18-1, 6 higher order bits select 1 of 64 circuits specific to the command. The remaining 2 lower-order bits assemble, on-the-fly, some combination of 3 reusable sub-circuits that any command might use. The figure below depicts this scenario:

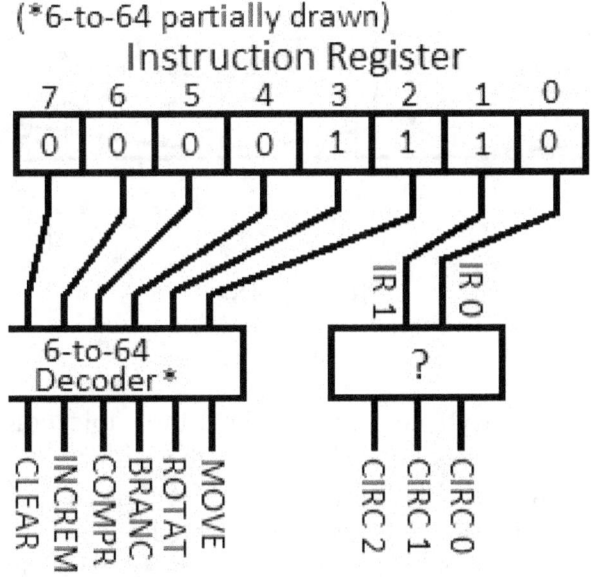

FIGURE 18-2

What makes the decoder labelled with a question mark (Figure 18-2) different from the decoders in Figure 18-1? The outputs of the "question mark" decoder can enable more than one circuit at the same instant of time. Each 2-of-4 decoder--seen in Figure 18-1--can enable only one circuit at a time. The new circuitry--seen inside the question mark encoder--though simple, can grow rather large. That's why I limited this one to being only able to select combinations of 3 sub-circuits.

Let's first make up a truth table, to see what we are asking the "?" encoder to do.

IR 1	IR 0	CIR 2	CIR 1	CIR 0
0	0	1	1	0
0	1	0	1	0
1	0	1	1	1
1	1	1	0	1

FIGURE 18-3

The IR 1 and IR 0 columns of the truth table contain the four possible pairs of values that can be in the rightmost 2 bits of the instruction register: 00, 01, 10, and 11. For any given row, a 1 in a CIR column means that CIR circuit is to be enabled.

So, for example, look at row 1. It says: if the last 2 bits of the command (IR 1 and IR 0) are 00, then sub-circuits CIR 2 and CIR 1 should be enabled (1), while sub-circuit CIR 0 could be disabled (0). Any command can build different sub-circuits by using different values of 1's and 0's in IR 1 and IR 0. A 1 enables; a 0 disables the sub-circuit.

We need to design three separate circuits: one for the CIR 2 output, one for the CIR 1 output, and one for the CIR 0 output. The procedure for each circuit is similar, so I will only explain how to design the one for CIR 0.

Look at the things circled in Figure 18-3. First, I have circled the entire CIR 0 column, because we are designing the CIR 0 enable circuit. Second, in the CIR 0 column, I have circled the 1's, because a 1 indicates when CIR 0 should be enabled. Third, in the rows where I have circled the 1's, I have circled the IR 1 and IR 0 inputs, because these are the input values that must trigger the CIR 0 circuitry to turn on. (They may also trigger other sub-circuits to turn on, but that will be part of the design for CIR 2 and CIR 1.)

Next, let's try putting into words what our design should do. Using the words, "and" or "or" is often inevitable, and leads us to the AND or OR gates we use. In the circled parts of Figure 18-3, we find that CIR 0 must be enabled (high) when IR 1 is high AND IR 0 is low, OR when IR 1 is high AND IR 0 is high.

Well, that was easy! The last sentence describes exactly the circuit we need. We input IR 1 and inverted IR 0 into an AND gate; then put IR 1 and IR 0 into another AND gate; then put the outputs of the two AND gates into an OR gate to produce CIR 0. This is called, "designing by truth table." You can see the CIR 0 enable circuit in the top third of the picture below. The circuits for CIR 1 and CIR 2 are designed directly from the truth table, just like CIR 0.

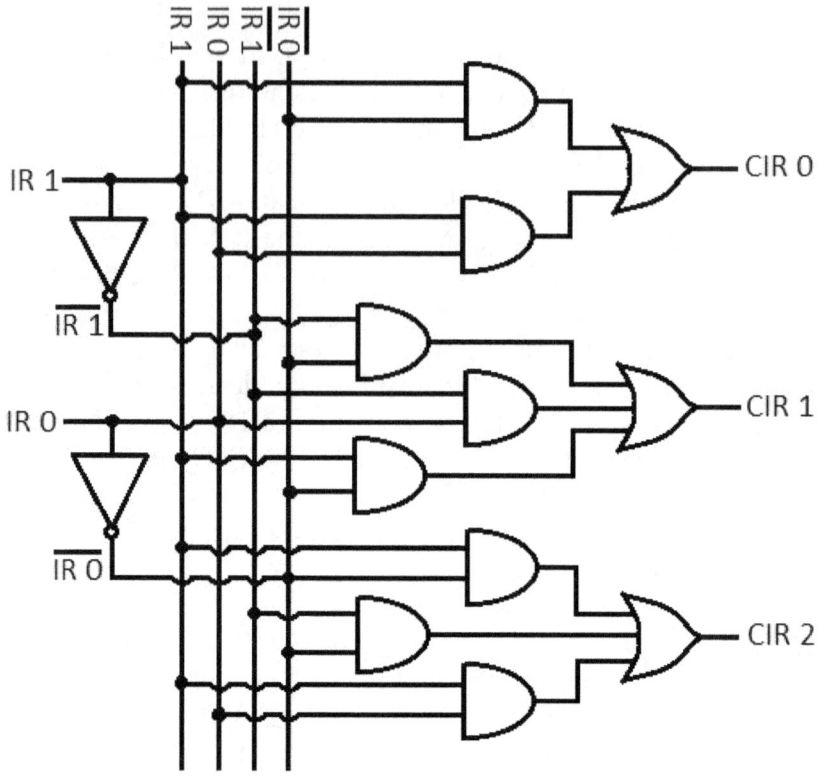

FIGURE 18-4

Designing by truth table is quick, and requires little thought. But, it often produces a circuit that is needlessly complex. There is a whole branch of study that reduces these resultant circuits into more compact forms that require less circuitry. These techniques have fancy names, like, "Karnaugh mapping" and "Boolean algebra simplification." In a way, they are fun, like solving puzzles.

However, for simple circuits like the one in Figures 18-3 and 18-4, a little extra thought is often all you need to simplify the circuit. The circuit below is a much simpler replacement for the circuit in Figure 18-4.

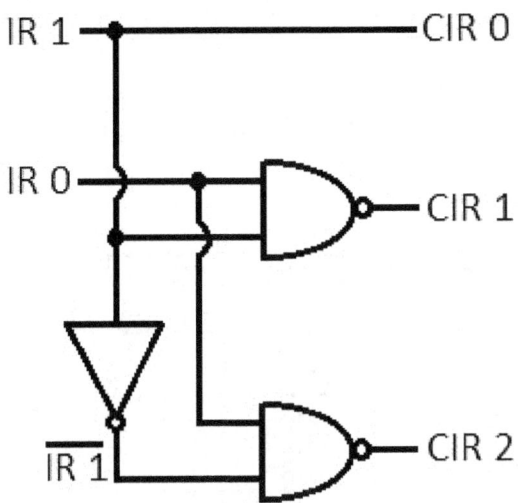

FIGURE 18-5

I derived this simpler circuit from close examination of the truth table in Figure 18-3. From that truth table, notice that to produce a high output out of CIR 0, the logic state of IR 0 doesn't matter; all that is required for CIR 0 to be high is for IR 1 to be high. Next, looking at the truth table's CIR 1 column, we see that it is the same as the NAND gate's truth table; so, all we really need for CIR 1 is a NAND gate. Finally, looking at the truth table's CIR 2 column, we see that the only time CIR 2 is low is when IR 0 is high and IR 1 is low; that leads to the CIR 2 circuit with the NAND gate and the inverted IR 1. The above circuit Figure 18-5, belongs in the box with the question mark in it, in Figure 18-2.

What if the selected sub-circuits chosen by CIR 0, CIR 1, CIR 2, and CIR 3 don't all interact simultaneously? That is, what if the results from CIR 3 must first be produced, before CIR 1 can operate upon it? What if CIR 2 must happen before CIR 1, which must happen before CIR 3?

This reminds us of our sequencer circuit, from Figure 17-1 (take a look at it, now). But, the demands on that sequencer were much less stringent. Using decimal, for simplicity, it only had to count in a very predictable order--0, 1, 2, 0, 1, 2, 0, 1, 2, etc.

We need a sequencer that counts in different orders. First, it may be asked to count 3, 1, 2, 0. Next, it may be asked to count 2, 3, 0, 1.

But we are being even more demanding of our new sequencer. We are asking it do counts of different lengths. First, we may ask it to count 3, 2, 1, 0. That's 4 counts. Next, we might ask it to count only 1, 0. That's 2 counts.

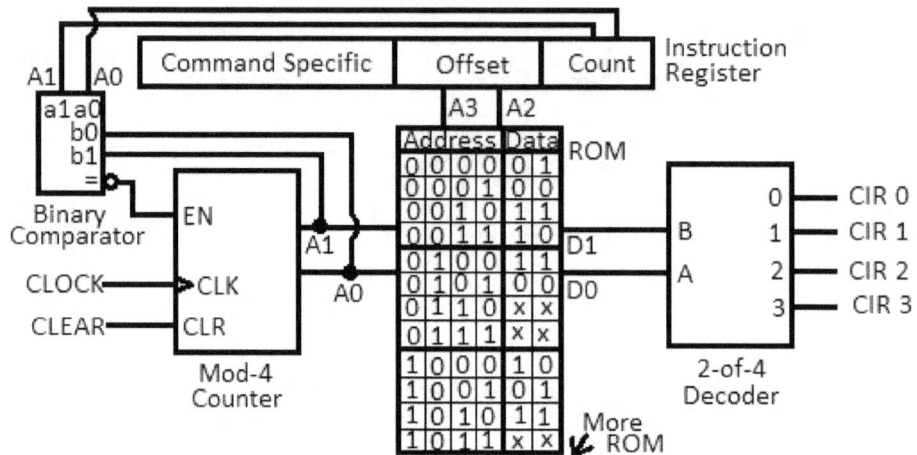

FIGURE 18-6

Figure 18-6, above, is one solution. In short, we have taken our sequencer from Figure 17-1 and placed a fixed-program ROM memory chip between the two parts. (The mod-3 counter is now a mod-4 counter.)

The ROM acts as a translation device, or as an encoder. The mod-4 counter still can only count in the 0, 1, 2, 3 pattern. It's the ROM that converts the pattern to whatever count sequence we want: 3, 1, 2, 0 or 0, 3, 2, 1, etc.

The ROM is in the center of Figure 18-6. The first 4 bits in each row is the address of the ROM location. The last 2 bits in each row are the contents of the ROM location. Those last 2 bits are what potentially come out of the ROM and activate the decoder, and hence the CIR sub-circuit.

I have marked off the ROM into groups of 4 rows. The 2 bits in the instruction register marked "offset" select which 4-row group is being chosen. Their A3 and A2 labels--for address 3 and address 2--mean that they are physically connected to ROM pins represented by the two leftmost binary numbers in each row: the two highest order address bits.

The two lowest order address bits--the third and fourth numbers in each row--are provided by the mod-4 counter. Thus, "offset," in the instruction register selects a 4-row ROM group, and the mod-4 counter moves though that group, one row at a time. Whatever contents the ROM designer has placed in the 2 bits of each row comes out, translated from the original 2 bits coming in on A1 and A0.

Let's do an example. Let's say the instruction contains a binary 00 in its "offset" bits. That means A3 and A2 of the ROM's address bits are both 0. Looking through the 2 leftmost columns of ROM, we find a match of 00 in the first 4 rows of ROM.

The mod-4 counter cycles through those first 4 rows of ROM. Its count sequence can be seen as the counts 00, 01, 10, and 11 in the 3rd and 4th columns of ROM.

Look now at the 5th and 6th column of rows 1 through 4. These bits contain the output sequence for our current example. When the mod-4 counter counts 00, the ROM outputs 01 to the 2-of-4 decoder, thereby "translating" the first count from 00 into 01.

Still focused on the first 4 rows, as the mod-4 counter counts 00, then 01, then 10, then 11, the ROM outputs 01, then 00, then 11, then 10, respectively. The 2-of-4 decoder, instead of sequencing through CIR 0, then CIR 1, then CIR 2, then CIR 3 without the ROM, now counts CIR 1, then CIR 0, then CIR 3, then CIR 2 with the ROM translator.

If "offset" in the instruction register contains a different value, a different translation occurs, hence a different sequence order out of CIR 0 through CIR 3. For example, if "offset" contains 01, then the second group of 4 rows is selected from ROM.

Notice that, in that second group of 4 rows, I have listed the translated count order as 11, then 00, then xx, and then xx. In binary listings, x stands for, "don't care." In Figure 18-6, we don't care about those xx memory locations, because they are never read. They live in 4-row ROM areas in which not all 4 CIR sub-circuits are enabled. In the second group of 4 ROM rows, only CIR 3 (from 11) and then CIR 0 (from 00) are enabled.

The mechanism to prevent these memory locations from being selected begins with the "count" bits in the instruction register. The "count" bits contain the binary representation of how many sub-circuits are selected (for that instruction) within the "offset" ROM area.

The "count" bits become the "A" inputs to a circuit new to us. It's called a binary comparator. The changing count bits of the mod-4 counter become the "B" inputs to the binary comparator. The binary comparator's enable output goes low when A1, A2 equals B1, B2. This low signal stops the mod-4 counter from counting beyond the limit imposed by "count" in the instruction register.

So, for the second group of 4 rows in ROM, "count" must have a 01 in it. This makes the mod-4 counter stop at a count of 01. Now, 01 is the 2nd binary number. And, 00 is the first binary number. So, CIR 3 is selected when A1-A0 is 00; CIR 0 is selected when A1-A0 is 01; then the execution of the command is done. The binary comparator stops the counter from going to the next count: A1-A0 equals 10.

Circuitry specific to the command would be responsible for clearing (CLR) the mod-4 counter to zero at the start of the command, and for enabling the clock input. Command-specific circuitry would, of course, also be responsible for tying together the sequential results from the CIR sub-circuits.

If it bothers you that this design wastes some memory locations in ROM (the xx areas), that problem can be fixed. We would make the "offset" area in the instruction register bigger, to include the total address (all 4 bits) of the beginning of the offset area into ROM. We would keep the "count" and binary comparator functionality unchanged.

However, we would change the way we create the address bits of the ROM. We would perform a binary addition of the "offset" value with mod-4 counter's changing values, to step though the ROM, with no wasted address space. I will show you a circuit design for a binary adder, shortly.

The microprocessor design using the circuit in Figure 18-6 might be used to accomplish more with one command than a microprocessor using the simpler designs in Figures 18-1 or 18-2. Microprocessor architectures designed with more complex circuits that accomplish more with each command are called CISC (complex instruction set computer) microprocessors. Microprocessor architectures designed with simpler circuits that do less with each command are called RISC (reduced instruction set computer) microprocessors. Both RISC and CISC styles accomplish the same things. The CISC microprocessors takes longer to perform one task. The RISC breaks the task down into more steps, but each step takes a shorter time. Some people feel that the RISC approach is the better of the two. *If* both tasks take the same time, the RISC advantage is less circuitry, saving on silicon space and power consumption.

What's the point of delving into all these circuits involved in decoding and executing microprocessor instructions? The point is, as always, to give you a feel for how the microprocessor works. We have done that on many different levels: how electronics works; how transistors work; how digital voltages are moved around, copied, and stored; etc.

Now, we have shown how microprocessors work on a much · deeper, broader, more complex level. Hopefully, you have been impressed by the power and flexibility of the microprocessor. You have seen that:

1.) Any one, specific command can be configured (Figure 18-1) in many ways; and each way rebuilds and reconnects circuits together to make a new electronic machine.

2.) There are many commands, each one with different powers, capabilities, and purposes; each one using different internal circuits.

3.) Putting commands together, sequentially, makes a program; thereby increasing the power of a single command, since the current command can build upon the result of previous commands.

4.) The microprocessor can run many programs, each one different, leading to innumerable combinations of internal circuits and countless tasks it can perform for us.

CHAPTER NINETEEN
THE BINARY ADDER

I mentioned the need for a binary adder, several paragraphs ago. A binary adder is a circuit that adds two binary numbers together, to produce their sum. For example, 1001 plus 0101 equals 1110.

I would not go into any detail about circuitry for a binary adder if the adder was simply used as part of a calculator application. But the fact is, the adder is used as an integral part of the basic operation of the microprocessor.

For example, the normal flow of a program is sequential. A command is fetched from a command memory address. Let's call that address "A". The next command is fetched from address $A+1$, because the program counter got clocked to the next count. We don't need an adder to increment by 1. A counter will do.

But, sometimes we want to get our next address from a different area of memory. Programmers know that "if" statements, "for" loops, and function calls cause the execution of code to be redirected to other parts of the program, instead of to the next line of code.

Often, a binary adder is used to calculate the address of this new, non-sequential, section of code. The contents of the program counter will be added to an offset number built into a command. The resulting sum will be put into the program counter. The next command will be fetched from this new, non-sequential location.

For example, say the program counter, pointing to address 68, fetches a command called BRANCH IF ACCUMULATOR EQUALS ZERO, 20. (Numbers in decimal, for simplicity.) When the fetch is done, the program counter increments by 1, becoming 69. Here's how this command works. If the value in the accumulator register is non-zero, the next fetch comes from address 69. But, if the value in the accumulator *is* zero, 20 (from the command) is added to 69 (from the instruction register). *A binary adder does this*. The sum, 89, goes back into the program counter, replacing the 69. Hence, the next command is fetched from a different area of code, at address 89, when the contents of the accumulator is zero.

Programmers will recognize the above description as the basis for the "if" statement. The "if" statement tests the validity of a comparison between two values. If the comparison is true, as in "does x equal 0?", then the next-in-line command takes place. If the statement not true, then a certain number of lines of code is skipped, so that a different command is executed.

The picture below shows the circuitry for a 1-bit binary half-adder:

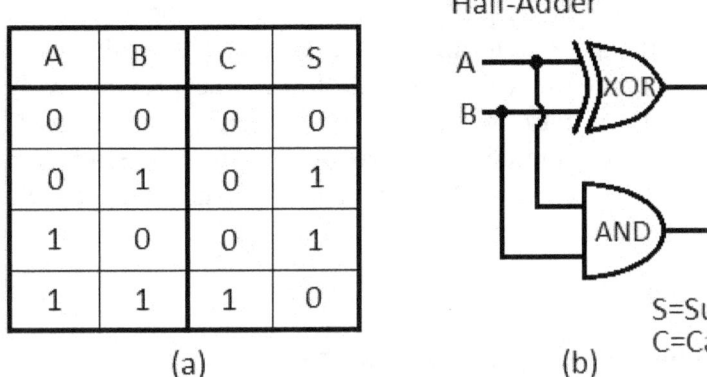

Half-Adder

A	B	C	S
0	0	0	0
0	1	0	1
1	0	0	1
1	1	1	0

(a)

S=Sum
C=Carry

(b)

FIGURE 19-1

The circuit above is called a half-adder because there is something incomplete about it: it does not accommodate a carry input bit. Binary, like decimal, adds numbers column-by-column, sending a carry to add to the next more significant column when the sum exceeds the base of the number system. (See Chapter Nine.) The half-adder is used for addition of the least significant bits of two binary numbers. For example, when adding binary $a_3a_2a_1a_0$ to $b_3b_2b_1b_0$, the half-adder handles the addition of b_0 to a_0, and provides a carry bit to the b_1 and a_1 column.

Once again, digital electronics amazes us with how easy it is to create the circuits we need. Of course, this presupposes all the work it took to learn basic electronics, transistors, and the circuits underlying gates and latches. Once we get comfortable with these, putting together circuits with gates and latches often seems natural and simple.

The above half-adder is a good example of this simplicity. Look at the truth table for the AND gate, in Figure 6-22, part(a); and for the XOR gate, in Figure 6-29, part(a), to understand how the half-adder circuit in Figure 19-1, part (b), produces the behavior shown in the half-adder's truth table of part(a).

In the truth table, you need to know that the S output is the sum output, and the C output is the carry output. Reviewing your basic binary addition, you should already know that 0+0=0, 0+1=1, and 1+0=0. All three of these additions produce a 0 for a carry bit. However, 1+1=0, with a 1 for a carry bit.

We need a slightly more complicated circuit for the binary addition of any bits with higher weight than the least significant binary bit. These circuits differ in that they must accommodate the carry-in, provided by the carry-out from the addition in the previous column. Look at Figure 19-1, part (b), above, to see the carry-out bit. That carry out bit must be handled by a carry-in circuit. This slightly more complicated circuit is called a full-adder, and is shown in the picture below:

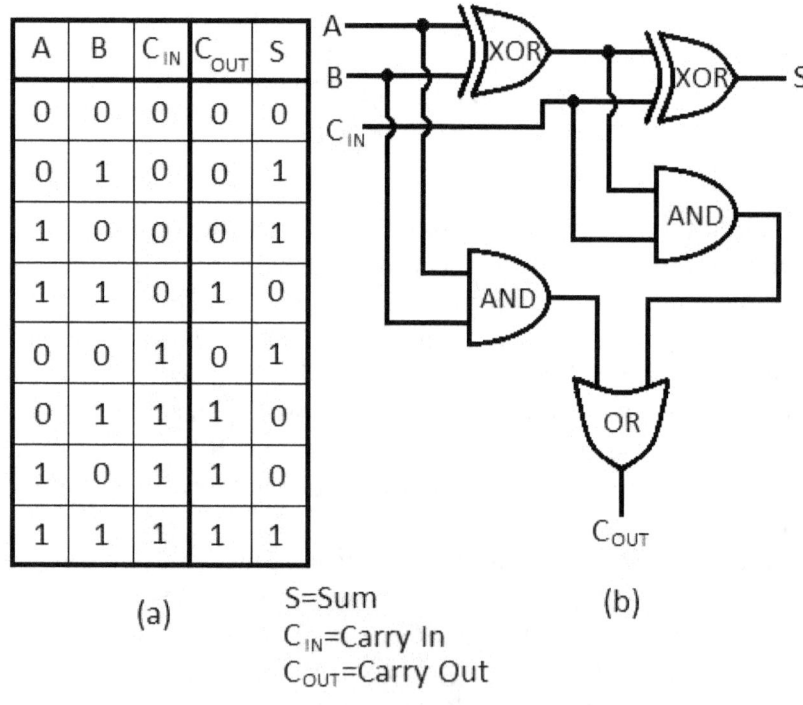

A	B	C_{IN}	C_{OUT}	S
0	0	0	0	0
0	1	0	0	1
1	0	0	0	1
1	1	0	1	0
0	0	1	0	1
0	1	1	1	0
1	0	1	1	0
1	1	1	1	1

(a)

(b)

S=Sum
C_{IN}=Carry In
C_{OUT}=Carry Out

Full-Adder

FIGURE 19-2

To understand how the full-adder implements its truth table, refer to the truth tables for the OR gate (Figure 4-16, part (a)), the AND gate (Figure 6-22, part (a)), and the XOR gate (Figure 6-29, part(a)). The full-adder's truth table and circuit now include Cin (carry input) to handle the carry output from the previous, less significant bit.

The first 7 rows of the full adder's truth table are not really different from the half adder. When an extra zero is part of a summation, it doesn't change the answer. So, the answer doesn't change if you add 0+1 compared to adding 0+0+1; you still get an answer of zero.

The 8th row is different, though. The answer is a sum of 1 and a carry out of 1, something we never got with the half-adder. We are adding 1 plus 1, plus a carry in of 1. 1 plus 1 gives a sum of 0 (a reset to 0) and a carry *out* of 1. But we then have to add the carry *in* of 1 to the old sum of 0, producing a new sum of 1.

Once you have memorized the truth tables for the basic gates, you can draw the above circuit on paper. Then, go line-by-line through the truth table, penciling in the three binary input values, A, B, and C_{IN}. Use your knowledge of truth tables to pencil in all the binary values in the circuit for each set of inputs. Prove to yourself that the circuit produces the correct output values, S and C_{OUT}, predicted by the truth table.

If you want to create a counter to add X bits together, simply chain X 1-bit adders together. For example, to make a 4-bit adder, chain four 1-bit adders together, as in the following picture:

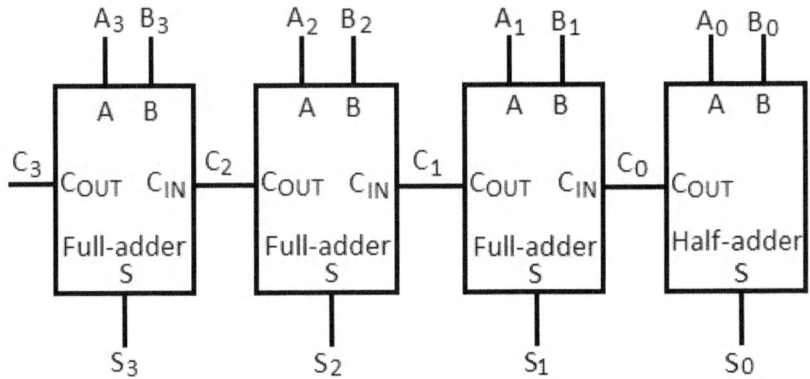

FIGURE 19-3

The adders are put together right-to-left, just like our number systems, with the least significant bit on the right, and the most significant bit on the left. As I mentioned before, the least significant bit can be a half-adder. The carries work their way from right to left; a carry out comes out of an adder and becomes the carry in for the adder on its left.

This type of adder is called a ripple-carry adder. Just be aware that there are adders that work faster, but are more complicated. One can also design a circuit that will either add or subtract.

CHAPTER TWENTY
THE STACK AND STACK POINTER

Twice we have made reference to the occasional need to change the normal behavior of the microprocessor, in its fetching of its next command sequentially. Remember, the microprocessor's program counter automatically bumps up by one count after a fetch. The number in it is ready to be placed on the address bus, to fetch the next command from the next higher address in program memory. We have noted that, after a testing type of command, the content of the program counter may be replaced with a different number, causing the next command to be fetched from elsewhere in program memory. We pointed out to programmers that this was the physical implementation of their "if" and "for" programming structures.

Sometimes though, the microprocessor needs the ability to return to the place where it stopped executing its sequential code. Say, for example (using decimal numbers for clarity), a fetch takes place from program memory address 25. The program counter bumps up automatically to 26. But the command from 25 executes, and replaces the contents of the program counter with 175. So, the next command is fetched from 175, then from 176, then from 177, then from 178, then from 179, then from 180, etc.

Sometimes, when the microprocessor is done executing the block of commands that started at address 175, it needs the ability to return where it left off, at address 26. Perhaps contained in memory location 180 is a command that says, "RETURN," which means, "Go back to where you left off." It won't contain the return address location of 26. It will just say "RETURN." The picture below illustrates this procedure:

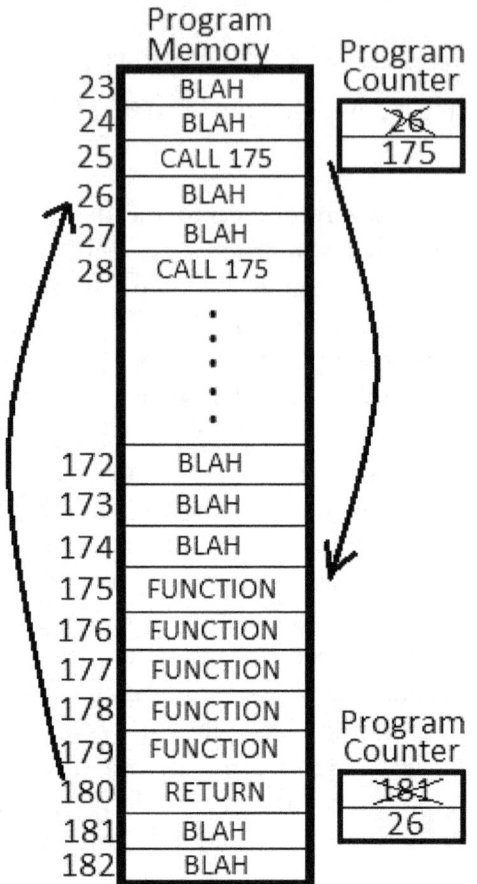

FIGURE 20-1

In the above figure, follow the arrows to see the program redirection and flow. Start at memory location 25, to the left of the program memory box. (Remember, there is only one program counter; don't let the figure fool you into thinking there are four of them.) The CALL 125 command replaces the 26 in the program counter with 175. Following the arrow on the right, program fetches are redirected to the 175-180 areas of program memory.

Next, look at memory location 180. The RETURN command

causes the program counter to be replaced with 26. Following the arrow on the left, program fetches are redirected back to address 26, in the main program thread.

Why would that command from address 25 make the microprocessor start fetching code from address 175, instead of from sequential 26? The block of code from 175 to 180 might be code that the programmer uses over and over, many times, throughout the program. (Notice in Figure 20-1 that CALL 175 is used a second time, at address 28.) Wouldn't it be great if the programmer could make one copy of the code and use it 50 times, instead of having 50 copies of the same code, scattered throughout the program?

The name for that "one copy" of repeatedly used code that lives at our address 175 through 180 is a "function." The name for the command at address 25, that redirects program execution to the function, is a "function call." The name for the command at address 180, that redirects the program execution back to the main program thread, is a "function return."

Let's calculate the efficiency of this technique in making the program more compact. First, let's look at the number of lines of code needed if we *don't* use functions. The repeated code, as found in the function is 5 lines, from 175 through 179. If we rewrote that code 50 times, it would cost us 5x50=250 lines of code.

Now, let's look at the number of lines of code needed if we *do* use functions. The function requires 6 lines of code, since it includes an extra line: the "RETURN" command. Each of the 50 times we call this command, we call it with 1 line of code, the function call. Now, it would cost us 50+6=56 lines of code.

Obviously, 56 lines of code using functions is much better than 250 lines of code without functions. Now, the same block of code may not necessarily be used as many as 50 times. But, a function is usually much bigger than 5 lines of code. It can easily be hundreds, if not thousands of lines of code. The efficiency of using functions becomes enormous.

You may also wonder how often a program would need to run the *exact* same code. If ever the programmer needs to do even one thing different from the function, he or she could nor re-use the function.

And, you would be right. Except, the function has built-in capabilities to overcome this shortcoming. The circuitry accommodates sending different binary values to the function each time the function is called. The circuitry also accommodates sending binary values from the function, back to the main program thread, when the "RETURN" command is called.

This added flexibility allows the function to run differently each time it is called. It can, for example, do a different calculation each time it is called, by sending it a different set of numbers to calculate. Each time, the function would send a different answer back to the main program thread.

There's still a mystery to be solved by the above description of the function call and return process. When the RETURN command executes, where did the 26 come from, to be placed in the program counter? We know where the 175 came from, that was put into the program counter to start the function; it was built into the command, CALL 175. But the RETURN command doesn't say, RETURN 26, it just says RETURN. It can't say RETURN 26, because it doesn't always return to 26. When we CALL 175 at address 28, it has to RETURN to 29, not to 26.

The answer to this mystery involves two new, related topics: the "stack" and the "stack pointer." The stack is a special area of RAM memory, used to temporarily store the address where your program left off when it got redirected to run the function. In our example, the stack stores the 26. When the function is done and RETURN is called, the "remembered" address in the stack, 26, is put back in the program counter, so your program can pick up where it left off in the main program thread.

This is an intuitive process. It's like what you would do if you had a job in which you were constantly being interrupted. You would write yourself little notes to help you remember where you left off, before the interruption. If you were about to do step 26 in your main task, you would write that down on a note, then do subtask A. When subtask A was done, you would refer to your note, to remember that you need to go back to the main task, step 26, to pick up where you left off.

But, what if, while you were in the middle of subtask A, you got interrupted again? You would write another note, saying what you were about to do next in subtask A. You would put the new note on top of the first note. (This is a "stack" of notes, get it?) Then, you would proceed to the new subtask B.

When subtask B is complete, you would refer to the top note on the stack, which would tell you where to pick up where you left off in subtask A. You would throw away that note (don't put it back on the stack), and finish subtask A.

Upon subtasks A's completion, you would refer to the note on the top of the stack. That note would tell you where to pick up where you left off in the main task.

The stack memory area in RAM is like a place to pile notes of where to return, to finish interrupted tasks. However, since an interrupting task can itself be interrupted, we need a mechanism to keep track of where we are on the stack. That mechanism is a register called the stack pointer. I will give an example of a stack that grows toward lower addresses, although stacks that grow toward higher addresses also exist. See the picture below:

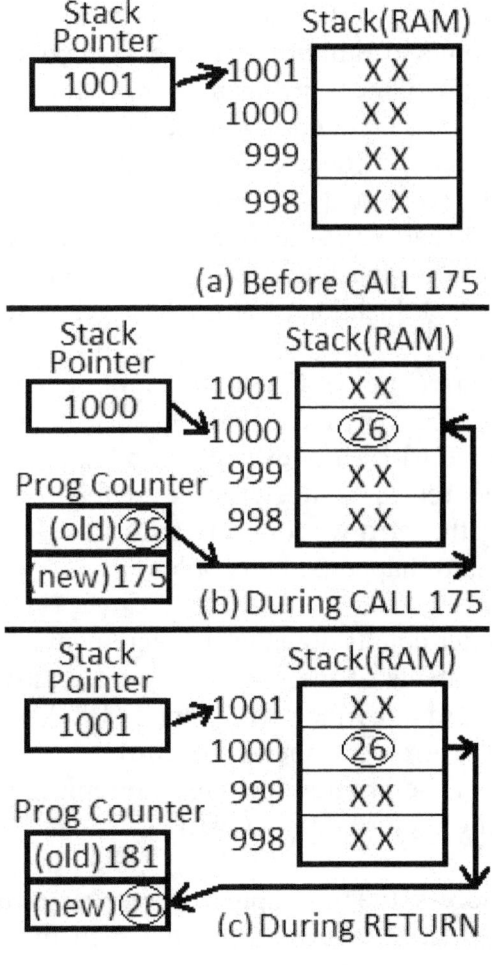

FIGURE 20-2

I have divided Figure 20-2 into three sections, part (a), part (b), and part (c). Let's first get comfortable looking at the figure. The stack pointer is a register--a set of flip-flops--arranged as a counter. Its contents are used to select the address of the stack area of RAM memory. In this, it is much like the program counter, with which you are already familiar. Also like the program counter, the stack pointer lives in the CPU (central processing unit) area of the microprocessor. The stack pointer differs in that it can count up or down.

I have drawn the stack memory on the right side of each section of Figure 20-2. The stack memory is a dedicated part of RAM, reserved for keeping track of return addresses. "XX" inside any stack location means we "don't care" what's inside. In each section of Figure 20-2, I have drawn an arrow from the stack pointer to the address in the stack it "points to". ("Points to" implies an address decoder, not shown.)

Finally, "Prog Counter" in each section refers to the program counter. The program counter is one register, not two. I show two boxes so you can see the contents of the program counter during (old) and after (new) the command.

Let's start with part (a). The stack is empty. We are in the main program. We haven't reached the CALL 175 command, yet. The stack pointer points to stack location 10001. The stack is full of "don't care" values.

Moving to part (b), the CALL 175 command is fetched from program memory 25. When the fetch is done, the program counter bumps up to 26. CALL 175 executes, and our mystery is solved. Here's where the return address, 26, is safely hidden. Follow the circled 26's and the arrows connecting them. During CALL 175's execution, three steps happen, in this order:

1.) The stack pointer counts down by 1 (decrements), to 1000.
2.) The current contents of the program counter (26) is written to the stack address pointed to by the stack pointer (1000). (Again, follow the circled 26's and the arrows connecting them.)
3.) The contents of the program counter are replaced with the address (175) of the function referred to in CALL 175.

26 is the address the microprocessor must return to when the function has finished execution. We have seen that it is now safely tucked away in the stack.

Now, let's see how the return process works, in part (c). The microprocessor fetches the RETURN command from the function's address, 180. When the fetch is done, the program counter bumps up to 181. During RETURN's execution, two steps happen, in this order:

1.) The contents (26) of the stack location pointed to by the stack pointer (1000) are written back into the program counter (Again, follow the circled 26's and the arrows connecting them.)
2.) The stack pointer counts up by 1 (increments), to 1001.

The stack is once again empty. Realize that if another CALL had happened before the RETURN, the stack would have grown down into address 999. Here's an example. Figures 20-1 and 20-2 will help you visualize this. Let's name the function at address 175 the "A" function. At address 177 of the "A" function, a call to a "B" function occurs. Address 178 is saved at stack address 999. When the "B" function finishes with a RETURN command, the 178 address is "popped" off the stack, and put in the program counter. Then, when the "A" function finishes with its RETURN command, the 26 address is popped off the stack, and put in the program counter. We are back in the main program, where we left off before the "nested" functions: the function within a function.

Do you see how much of the microprocessor's behavior is based on moving information around? Remember back in the beginning of this book how I stressed how amazing and important it is that a group of binary voltages can be moved from one location to another?

With this in mind, let's take a moment to appreciate the ballet-like beauty of the above process. Think of the program counter changing, commands moving, the stack pointer changing, addresses moving in and out of the stack. All of this accomplished with precision and a kind of beauty, and at the microscopic scale. It's like the beauty of the internal combustion engine.

CHAPTER TWENTY-ONE
INTERRUPTS

There's another, similar situation in which the above stack and stack pointer mechanism is used. Above, we covered the need to remember and return to an address when a function is called. Sometimes, the microprocessor must remember and return to an address when a "hardware interrupt" occurs.

Here's the basic concept behind a hardware interrupt. The microprocessor is surrounded by subsystems that may need the microprocessor's attention. A person may be pressing a key on a keyboard. The USB port may have just received new position information from the mouse. The Ethernet port may have just received the web page you requested. The printer may be ready for another page of the document you want to print. The hard drive buffer may be ready with the text document you want to edit. There are so many things that need the microprocessor's attention.

Now, the microprocessor could sit in a main program loop, and constantly check one subsystem at a time, to see if any of them needs attention. This technique is called polling. It is a software solution. But, this solution wastes time. Since it's a software solution, all of this polling wastes time in fetching, decoding, and executing. It's a waste when a subsystem is in no need of the attention. Plus, the subsystem that does need attention has to wait while the microprocessor checks subsystems that don't need attention.

Hardware interrupts are a better, more efficient, solution. Think of a teacher in a classroom. The teacher gives an assignment that may require individual help. The teacher could use the polling system, and walk around to each student, one at a time, and say, "do you need help?" A "yes" answer would get the teacher's help. A "no" answer would be a waste of the teacher's time. The student that truly needs help will have to wait until the teacher checks with every other student, who may need no help.

Obviously, the better solution is a hardware interrupt. The teacher tells the students, "whoever needs my help should raise his or her hand."

The hand is the hardware. It is binary: a raised hand is true, a not-raised hand is false. The teacher does not waste time this way. The first student to raise a hand gets immediate help.

In a microprocessor, hardware interrupts are handled by specialized circuitry in the CPU. The basic idea is, that when a subsystem needs the microprocessor's attention, it pulls its interrupt line (hardware) high. The microprocessor:

1.) finishes its current command
2.) saves the address it must come back to on the stack
3.) places the address of the subsystem-servicing code in the program counter
4.) runs the subsystem-servicing code until it encounters a RETURN command
5.) restores the return address from the stack to the program counter, resuming the code "interrupted" by the hardware interrupt

A hardware interrupt is even more amazing than a function call. A function call has a little bit of predictability. Though the microprocessor can't anticipate when a function call will occur, the programmer can. The programmer puts the code (software) exactly where he or she wants it.

Neither the microprocessor nor the programmer know when the hardware interrupt will occur. This has major implications. It means that neither the microprocessor nor the programmer knows what line of code is being executed when the interrupt occurs.

Look at the function calls in Figure 20-1. We know they happen when the code at line 25 and line 28 are fetched. We know the microprocessor must return to lines 26 and 29, respectively, when the functions are finished running.

Perhaps we could have designed a simpler mechanism than the stack/stack pointer, if we had this amount of predictability. But hardware interrupts occur unpredictably. We can't plan for them, in code. We don't know, in advance, what line of code will be executing when a hardware interrupt occurs. We really need the stack/stack pointer mechanism to handle them.

If you re-read the above 5 steps describing how the microprocessor handles an interrupt, you realize that it's similar to how it handles a function call and return. The big difference comes in step three: "The microprocessor places the address of the subsystem-servicing code in the program counter."

With a function call, the address of the function-handling code is built right into the CALL command. If the command is CALL 175, 175 goes into the program counter, to be executed next.

Things are not so easy with interrupts. We have to put into the program counter the address of the code that handles the interrupt for the particular subsystem. But, there are many subsystems, each with its own interrupt-handling code, each stored at a different starting address. A new mechanism is required, one that allows the correct address to be put in the program counter, to access the interrupt-handling code needed by the interrupting subsystem.

For example, let's assume there are 10 subsystems, labelled A through J, each with the ability to interrupt the microprocessor. Each of the 10 subsystems gets serviced by its own unique code, different from the other 9, and stored at a different starting address from the other 9. Assume these code areas are also labelled 0 through 9. We need a mechanism so that, for example, when subsystem 2 interrupts the microprocessor, only the code in code area 2 runs. Only the starting address for code area 2 gets put in the program counter. The picture below describes this scenario:

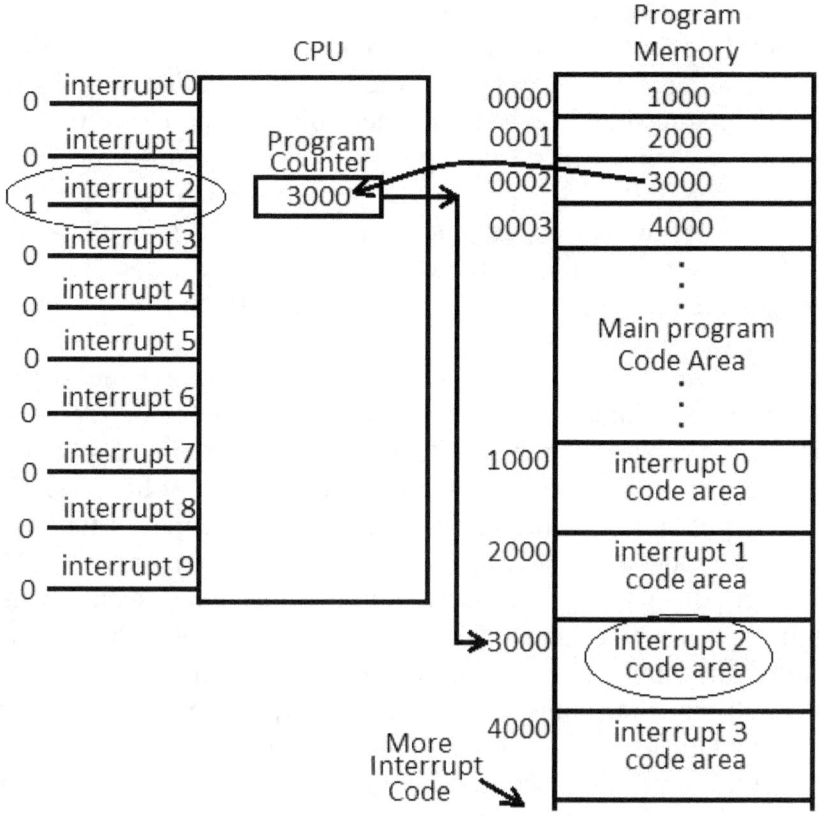

FIGURE 21-1

In this figure, you see that only the interrupt for subsystem 2 is true (logical 1, or 5 volts input on the left side of Figure 21-1). The code area to service the interrupt 2 subsystem begins at address 3000, and is being activated, because the program counter has been loaded with address 3000.

Different microprocessors use different designs to put the correct address for the interrupt in the program counter. In Figure 21-1, I have shown one such technique. An area of memory is reserved for use as an interrupt vector table. The address for interrupt 0's code area is stored at the first address in the table; the address for interrupt 1's code area is stored at the second address in the table; etc. (When you first see something like this, it can be disturbing, because these are addresses being stored as data in memory.)

You can find this table in the upper right side of Figure 21-1. In our example, interrupt 2 occurs. After saving the return address on the stack, built-in circuitry automatically reads the contents of address 0002 in the interrupt vector table. Address 2, remember, holds the address of the interrupt 2 code area. The built-in circuitry then writes the 3000 value it found there to the program counter. Code proceeds to run from address 3000.

When the code to service interrupt 2 is done, it ends with a RETURN command, similar to the way a function ends. I don't need to go over how the return address is stored to and restored from the stack. It's the same as for a function.

As I have mentioned, there are other designs for handling the generation of the interrupt address. In one technique, there is no interrupt vector table. Or, perhaps it's better to say the interrupt vector table consists of one address. No matter which interrupt happens, that one address gets put into the program counter. That means that only one section of code starts running, no matter which interrupt occurred.

However, once that code starts running, that code spends some time figuring out which interrupt occurred. It checks the voltage levels of the interrupt input lines, looking for a 1 (true). After it figures that out, it replaces the contents of the program counter with the appropriate address to service the correct interrupt.

As you can see, the second solution to generating the interrupt-servicing address is software-driven. A third solution is hardware-oriented. In this technique, special hardware built into each subsystem sends the interrupt-servicing address to the CPU. This event occurs after the microprocessor acknowledges the interrupt signal from that subsystem.

And, interrupts can be even more complicated and interesting. Some microprocessors prioritize their interrupts. Think of a CEO, the head of a big business. The CEO is like the CPU. The CEO doesn't mind being interrupted, but some interruptions are less important than others. Interruptions from the CFO (chief financial officer) might have top priority, because ignoring the CFO could lead to financial disaster. Interruptions from the CIO (chief information officer) might have second highest priority, because ignoring the CIO could lead to disasters from security breaches, communication loss, or data loss. Interruptions from the Marketing Director, on the other hand, might be less time-critical, and hence have a lower priority.

The CEO might give his/her secretary instructions about the priority of interruptions: a priority order:

A.) "If I am talking to the CFO (interrupt level 0), don't let anyone interrupt us."
B.) "If I am talking to the CIO (interrupt level 1), *don't* let the Marketing Director (interrupt level 2) interrupt us, but *do* let the CFO interrupt us."
C.) "If I am talking to the Marketing Director (interrupt level 2), let the CFO or the CIO interrupt us, but don't let anyone else interrupt us."

Prioritizing interrupts works similar to nested function calls, which we have already covered. That's where, during the executing of code for function A, a call for another function (function B) occurs.

To demonstrate a prioritized interrupt system, assume that interrupt-servicing code for subsystem A is running. Assume an interrupt request occurs from subsystem B. One of two possible things happen next:

1.) Subsystem B's interrupt priority is equal to or less than subsystem A's. In this case, the microprocessor will finish *all* the code for subsystem A before running the code for subsystem B.
2.) Subsystem B's interrupt priority is greater than subsystem A's. In this case, the microprocessor will finish only the current command for subsystem A, save the return address on the stack, then immediately go to work on the interrupt-servicing code for higher-priority subsystem B. When done with that, the microprocessor will return to completing the rest of the code for lower-priority subsystem A.

Looking back at Figure 21-1, the 10 interrupts could be prioritized so that interrupt 0 is of the highest priority, and interrupt 9 is of the lowest priority. (Strangely, it's usually backwards, like this.) Of course, with prioritized interrupt capability, more than one interrupt input can be at a logic 1 level (true), and the microprocessor would have no problem handling it. Digital circuitry would enable it to follow the rules outlined in steps 1 and 2, above.

You've come a long way, in your quest to learn how microprocessors work. We have moved from tiny electrons to less tiny transistors; one transistor to combinations of transistors; from simple circuits to complex. I hope that you have enjoyed and profited from this journey. And, I hope you can now look upon your microprocessor-controlled devices with even greater appreciation, now that you understand how they work.

APPENDIX A
CODE

Machine Language (Appendix A-1)

I suspect that you may wonder how we make this wonderful device do things for us. You have seen the pieces of the puzzle that fit together to make a microprocessor. You have seen how they work together. But, you may wonder, how does the code get inside these things, to make them work? How does a person go from buying a microprocessor, to creating a product with it that, say, vacuums my floors when I'm at work?

The process starts with the person or persons who design and create a microprocessor. Their criteria is based on the target market for the microprocessor. Does it have to be small and inexpensive, for simple tasks? Or must it be powerful, to run a computer server?

Whoever designs the hardware of the microprocessor must also design its command code. Let's assume a simple microprocessor, based on 8-bit code. The designer is immediately limited to $2^8=256$ binary commands. The first command is 00000000, the second is 00000001, followed by 00000010, etc., with the last command of 11111111.

The designers must decide what each command does. That is, they must decide how each command executes. For example, the command 00000000 might execute by resetting the contents of the accumulator to all zeroes. The command 00000001 might move the contents of the accumulator to the stack (a good way to send a number to a function).

If you think this through, you will realize that another team designing for another company might create a microprocessor that executes the command 00000000 in a completely different way, such as disabling all interrupts. Their command to reset the contents of the accumulator may be 11111111. This has important ramifications, which we will address later.

Part of the design process is to create a memory map to determine the addresses of the subsystems. The circuitry--like the address decoders that we have covered--must implement the memory map so that the correct subsystems are accessed by the commands. Importantly, when the fetch process occurs, commands had better be coming out of command memory.

Assume next that the circuitry is designed and built to carry out the decoding of the 256 chosen commands. The next step is to write a program, then put it into command memory.

The first home computer was the Altair 8800. It used the Intel 8080 as its microprocessor. It used RAM to store programs. RAM is volatile. "Volatile" means the program goes away when the device is turned off. Look at the simplified (reduced to 8-bit) picture of an Altair 8800, below:

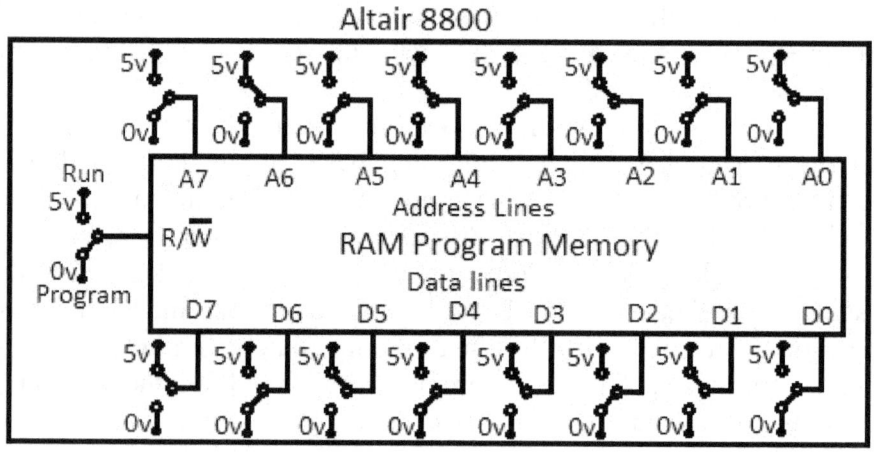

FIGURE A-1

The first Altairs required the user to put the command code into RAM program memory using switches. Switches on the outside of the box connected to address and data pins on the RAM inside the box. You flipped all the address switches to 0 volts, to select address 00000000; then set the data switches in the pattern of your first command; then flipped the Run/Program switch to 0 volts for a moment, to store the command at that address. Next, you raised the A0 switch to 5 volts, to select address 00000001; then set the data switches in the pattern of your second command; then flipped the

Run/Program switch to 0 volts for a moment, to store the command at that address. You repeated this process until your whole program was loaded. You finally could flip another switch to pull your microprocessor out of reset, so your program could run.

I mention this process because I think tracing this historical process helps your understanding of how you get binary code into command memory. No matter how the task is accomplished, the basic idea is as simple as the one shown above. You already know how switches work. Flip a switch up, in Figure A-1, and 5 volts appears on a RAM input line. Flip a switch down, and 0 volts appears.

Plus, you already know that we can replace the switches in Figure A-1 with transistors. You also know that these transistors can be built into circuits that can automate the task of loading a program into memory.

And, of course, that is historically what happened. The Altair 8800 switch-loading process was obviously intolerably slow. The next step in the evolution of getting command code into memory was the invention of the non-volatile ROM memory. This integrated circuit was memory that was programmed once with command code, then inserted in the circuit where the microprocessor could read it. Since it was non-volatile, the code stayed there when power was turned off.

Until relatively recently, if you needed to upgrade the ROM BIOS (the start-up command code) of the PC personal computers, you had to remove the ROM from the computer and replace it. With the invention of Flash memory, you can reprogram this hybrid RAM/ROM IC where it lives, in the PC. Understand that some microprocessors have the Flash memory for command code built right into the body of the microprocessor.

There is a second problem with the Altair's switch-loading process to put command code into program memory. In that process, you have to enter the commands in binary. In our above example, if you wanted to enter a command to reset the accumulator to all zeroes, you had to know that 00000000 was the command to enter. You had to either memorize the binary code for every command, or constantly look it up in a reference manual.

This native command language of microprocessors of 1's and 0's (of 5 volts and 0 volts) is called *machine language*. It is obviously too clumsy to work with.

Hence, the designer of a microprocessor will provide a related command language, called *assembly language*. Assembly language is a short, human-language description of what each of the binary machine language code commands does. There is a one-to-one correspondence between machine language and assembly language. If there are 256 machine language commands, there are 256 assembly language commands.

As an example, assume the machine language command 00000000 executes by resetting every bit in the accumulator to zero. The assembly language command for this might be: CLR ACC. This would be a good choice, because it is short for "clear the accumulator," which is short for "make every bit of the general purpose register named the accumulator become zero volts."

A person can program with assembly language much more easily than with machine language. It's easier to memorize, and easier to type without making mistakes than binary, especially if the size of the commands are 16 bits, 32 bits or 64 bits per command.

But wait: I just used the word "type." How do we go from flipping switches on the Altair to typing commands? Well, remember when I said we could automate the task of flipping switches? We automate the task of getting command code into memory by using another computer.

To program in assembly language, someone must write a computer program called an *assembler*. The assembler program has the ability to read a saved text file written in assembly language, change it into machine language, then save it as a new, machine language file.

Let's elaborate on this. An assembly language programmer types a line of code into a text editor program, like Notepad. Assume Notepad codifies text using ASCII code. For our CLR ACC command, the C in CLR gets saved in the text file as the ASCII code 01100111. The L in CLR gets saved in the text file as the ASCII code 01110110. This process continues until the whole program is finished and saved. Let's say it is saved as, "myAssem.src."

Next, the assembly language programmer runs the assembler program. The assembler needs to know which program to assemble.

The programmer enters, "myAssem.src." The assembler program "assembles" "myAssem.src." That means it converts the ASCII-encoded assembly language code into ASCII-encoded machine language code. It saves this as a new file: "myAssembly.obj."

Notice I said it saves "myAssembly.obj." in ASCII code. So, when it converts CLR ACC to 00000000 (which is 8 zeroes), it doesn't really convert it to 8 binary voltages. Since it is saving it to a text file, it saves 00000000 in the text file's alpha-numeric encoding format, which we have assumed is ASCII code. The ASCII code for zero is 01001000. The assembler saves 01001000 eight times, to save the binary command 00000000. (More frequently, to reduce file size, it saves the hexadecimal instead of binary version of the machine language code.)

"myAssembly.obj," the new machine language version our program, contains more information than just the command code. It also contains the address where each command will live in command memory.

The reason "myAssembly.obj" is not saved in binary is that we have one more step to accomplish in our task. We used a computer to assemble the code. Now, we use the computer to "burn" the code to the program memory.

First, we need a specialized hardware device called a programmer, or burner, designed to put the code into the program memory. This device connects to the computer and to the target memory to be programmed. I will call it a "burner" because calling it a programmer makes us think we are talking about a human software programmer. ("Burner" is an old term, from back in the days when ROM code was literally burned in by melting metal links in the IC's. I will call it a burner, even when referring to programming Flash memory.)

We also need a computer program that can take the assembled program, "myAssembly.obj," and send it to the burner. Unfortunately, this software program is also referred to as a programmer. I will call it burner software, because calling it a programmer makes us think we are talking about a human software programmer.

So, the burner software takes the "myAssembly.obj" code and sends it to the burner hardware. The burner hardware reads the code and converts it into true binary voltage levels of 5 volts (1's) and 0 volts (0's). Finally, it puts the code into the correct addresses inside the memory chip it is burning.

High-level Languages (Appendix A-3)

Even before microprocessors were invented, the creators of machine language and assembly language realized that programming a computer this way would severely limit the number of people willing to create programs this way. Even assembly language requires a deep knowledge of the underlying hardware (circuits) in the microprocessor. Assembly language programmers must be familiar with register names, interrupts, the stack, addresses, and different ways to generate addresses. Plus, if you work with a different microprocessor, command names, terminology, and techniques change.

So, they invented *high-level languages*. These languages have names like Fortran, COBOL, BASIC, C, C++, and Java. They are called, "high-level," not because they are harder to learn or more powerful. Assembly language is harder to learn and more powerful. They are called "high-level" because they are a step up from having to work at the hardware level. They allow a person to write a program without having to know how hardware or circuits work, without needing to know names of internal registers or their addresses. Plus, if a high-level language is standardized (meaning the commands are the same no matter who provides that language), you should be able to learn it once, and use it in any microprocessor-based system.

A high-level language is very different from assembly language. Primarily, it abandons the reliance on the one-to-one nature of assembly language to machine language. The goal is still to create a machine language program. But one command in a high-level language becomes many commands in machine language. That one command no longer resembles machine language commands. Gone are references to registers, addresses, etc.

For example, let's look at a java language "if" statement:

```
if(x>3){y=0;}
```

Java is a high-level language. The above code means, "If the contents of variable x is greater than three, then change the contents of variable y to zero. But, if the contents of variable x is not greater than three, then leave the contents of variable y alone."

This one line of java code, when reduced to assembly language, becomes many lines of code. The assembly language code references registers and addresses, and involves a knowledge of signed binary arithmetic.

The good news here is that high-level languages reduce to machine language in much the same way that assembly language programs reduce to machine language. The difference is, a "compiler" is the name of the software that converts a high-level language to machine language. (I won't discuss "interpreters," to avoid needless confusion.)

Small-system Code vs. Application Code (Appendix A-4)

We have reached a point where we have explained the earlier questions we posed. The above paragraphs explain how you can put code into program memory. They also explain how you can go from buying a microprocessor to creating a microprocessor-based product. Briefly, you:

1.) Buy the microprocessor
2.) Study its architecture, address space, and code
3.) Build circuits around the microprocessor with which it "interfaces"
4.) Write command code for the microprocessor
5.) Use the burner hardware and software to put the code into program memory
6.) Make lots of money

This description implies knowledge of electronics design. You don't have to design the microprocessor from the ground up, but you must know how it works, and how to build upon it, even if you are using a high-level language for step four. It helps to have a background in computer engineering.

That being said, just because you are not a computer engineer, that doesn't mean you are automatically left out of all the fun. There is another path to working with microprocessors that doesn't require intimate knowledge of how they work: software engineering. You may know software engineers--computer programmers--who don't know how microprocessors work. You may be one, yourself.

Software engineers work on microprocessor-based systems in which the above five steps are already done. All of the hardware design is complete. The product is already functioning. It already has some basic, burned-in code. It is a complex system, called a computer, or tablet, or cell phone.

In this kind of environment, a programmer adds more code to the code that already runs the computer. This new code extends the capabilities of the computer. Let's call this new code an application, or "app."

Writing an app does not require extensive knowledge of the

microprocessor or electronic hardware. All that work is taken care of by built-in code called the BIOS and by the operating system (I'll define the operating system, shortly). Designers of the operating system provide programmers with a "Software Development Kit", so the programmer's app code can connect with the operating system's functions. The BIOS and the operating system's functions handle the "dirty work" of controlling the physical electronics.

This kind of programming does not involve burners that put code into ROM. In a computer, an app is fetched from DRAM (dynamic RAM). The ability for larger systems to execute code from DRAM brings us to a whole new programming and hardware environment. The figure below helps explain where command code comes from, in a PC-style computer:

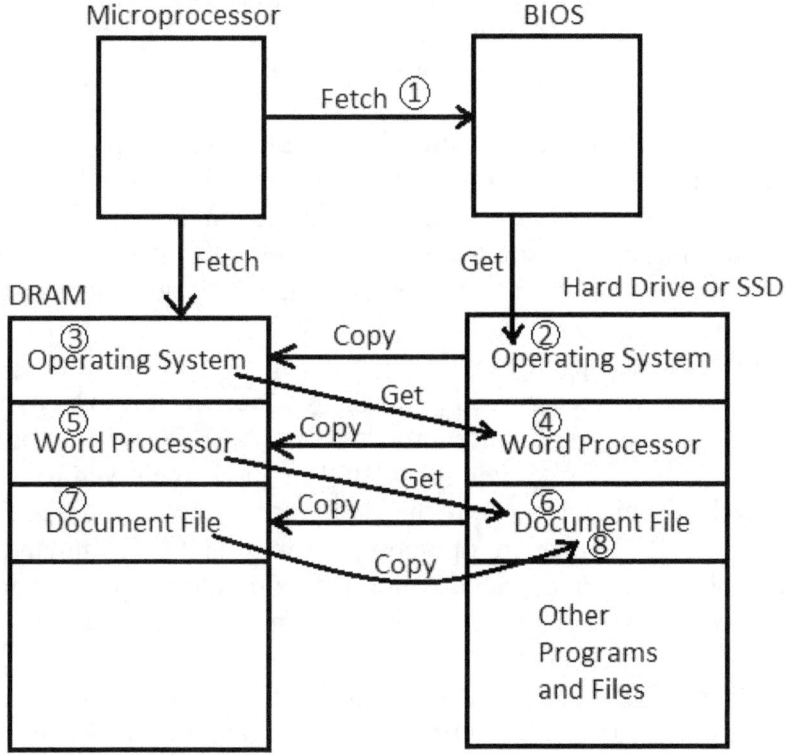

FIGURE A-2

Figure A-2 shows the large microprocessor-based system we think of when we talk about a "computer." In this system, the microprocessor normally fetches its command code from DRAM. DRAM is cheap; and it's large enough and fast enough for most purposes. However, it is volatile: it loses information when turned off. So, programs can't be stored in DRAM when power is off. It is empty when turned on. Furthermore, even if it was non-volatile, its capacity isn't great enough to store all the apps and files a modern computer user requires.

So, we need a massively large storage area that is non-volatile; one that retains apps and files when the unit is turned off. That's the job of the hard drive. (Or, solid state drive: flash memory for smaller, lightweight units like notebook computers.)

The problem is, the microprocessor doesn't fetch command code directly from the hard drive. The hard drive is too slow. It's an electro-mechanical device. It accesses information using motors. To a microprocessor, waiting for motor to move is like us watching grass grow.

So, code stored in the hard drive that the microprocessor must execute must first be moved to DRAM, then fetched from DRAM.

Follow the arrows and the numbers to understand Figure A-2. When the microprocessor wakes up at power-up, it must fetch code from whichever address it has been designed to retrieve. The computer creators put BIOS in that address space (step 1).

BIOS (Basic Input/Output System) is in a small memory space. It is just like the permanent code loaded by a burner, that we covered in our discussion of small microprocessor-based systems. It is a small, preliminary operating system. The BIOS is the name given to both the integrated circuit chip and the software (firmware, since it is permanent) inside it. An operating system is the command code responsible for the overhead of running a computer: loading programs, saving programs, maintaining a file system, input and output of data, etc.

BIOS performs some self-tests on the computer. It checks settings stored in the Setup IC. In Setup, it finds the boot order: the prioritized list of where to find the primary operating system, like Windows or Linux. In our example, the BIOS finds the primary operating system in the hard drive.

Still fetching from BIOS code, the microprocessor copies the primary operating system from the hard drive into DRAM (step 2 to 3).

When that is done, the code last code executed by BIOs changes the program counter to point to the address of the first command in the primary operating system, in DRAM. When the user uses an input device to load an app, say a word processor app, the operating system responds to that input by copying the word processor app from the hard drive to DRAM (step 3 to 4 to 5). The operating system lets some of the code fetches to be from the word processing app in DRAM. If the user of the word processing app loads an old document to edit, the operating system copies that document data from the hard drive into DRAM (step 5 to 6 to 7). Any changes made to the document are made in the DRAM copy. When the user wants to save the newly edited document, the operating system code copies it from DRAM back to the hard drive (step 7 to 8).

Thus, when a programmer creates an app and you install it in your computer, you are putting it into the hard drive. There, it becomes one of many other programs. When it is needed, it gets loaded into DRAM. From there, the code is fetched, to be decoded inside the microprocessor's CPU.

As you have seen, the answer to our question, "How do you get code into program memory," is not a simple one. Part of the problem is that there are different kinds of microprocessor-based systems. Some are simple and small. Others are complex and elaborate.

So, for example, for a small, simple microprocessor, code may simply need to be burned into ROM. A person may choose to program it in assembly language. People may do this because they love how microprocessors work; programming in assembly language makes them feel closer to the "workings" of the machine, to the "machinery" of the machine. Or, they may believe that assembly language programs are more efficient and compact. People did, at one time, also create assembly language programs for PC-style home computers. A program named "Debug" was used for this purpose.

On the other hand, in a large, complex microprocessor system like a computer, some code must still be burned into ROM. But, much more code--written in a higher-level language--can be stored in a hard drive but run from DRAM.

Write Once, Run Anywhere (Appendix A-5)

Now that we have discussed large microprocessor-based systems--computers--let's go back to my earlier reference about the following problem. You have microprocessor A, made by Atmel; microprocessor B, made by Intel; and microprocessor C, made by Microchip. You have created high-level language code.

However, the machine language code for each of these is totally different. You can't just run one compiler one time to change your high-level language code into machine language for all three microprocessors. The machine language produced for microprocessor A will not work for microprocessors B or C. You would have to run a high-level language-to-microprocessor A compiler; then a high-level language-to-microprocessor B compiler, then a high-level language-to-microprocessor C compiler.

That's why, traditionally, code written for PC-style computers did not work in Apple computers, and vice versa. PCs traditionally used Intel microprocessors, while Apples traditionally used Motorola microprocessors.

The Java language made a breakthrough with this problem, with a solution referred to as, "write once, run anywhere." With java, you compile your program once. The resulting code is partly compiled. It is compiled into an intermediate state, called "byte code." Then, you install a java virtual machine on your computer. This java virtual machine is software made specifically for your microprocessor. When you run your java program, the java virtual machine converts the byte code into the machine language that your microprocessor was designed to execute. Thus, you can distribute the same bytecode on the internet to any computer run by any microprocessor and, assuming it has the correct java virtual machine installed, it will execute properly.

APPENDIX B
D/A CONVERSION

As another bonus, I'd now like to take you one step beyond your knowledge of how microprocessors work. I have shown you how microprocessors interface with other digital devices. Now, I'd like to show you how microprocessors interface with analog devices.

This ability to interface with analog devices is important, because the "real" world, the human world, is analog. The real world consists of temperature, light, velocity, color, sound, wind, and force, to name a few. Each of these can be in countless states. A car can move at 10 miles per hour. Also, it can move at 15, or 20.5, or 30.22, or 30.222, or 30.2222 miles per hour. This quality of being in many measureable states is called *analog*. The quality of being in only two measureable states is called *digital*.

The ability of microprocessors to interface with the analog world is exciting, because it greatly increases their power and usefulness to us. It enables microprocessors to be able to monitor, and to some extent control, the real, analog world of temperature, light, velocity, color, sound, and force.

Before we had microprocessors, we had devices called *transducers*. Transducers are devices that change one form of energy into another. Some transducers change one aspect of our analog word into voltage. For example, a microphone changes a sound into voltage. A thermocouple changes heat into a voltage. A tachometer changes rotational velocity into voltage.

Other transducers work in the opposite direction. They change voltage into an element of our analog world. A heating element changes voltage into heat. An LED changes voltage into light. A speaker changes voltage into sound.

However, the voltage, in each of the above two types of transducers, is analog voltage. The microprocessor is a voltage-based device, but its world is one of digital voltages. What the microprocessor needs is one type of device that converts analog voltages into digital voltages: an A-to-D converter (analog-to-digital). And, it needs another type of device that converts digital voltages into analog voltages: a D-to-A converter (digital-to-analog).

To state this more accurately, the microprocessor needs a device that converts one analog voltage into a group of digital voltages. Then it would be able to sample the real world. The real world first gets turned--by a transducer--into an analog voltage, which gets converted to a group of binary voltage by an A-to-D converter, which gets input by the microprocessor. (See part (a) of Figure B-1, below.)

The microprocessor needs a second type of device that converts a group of binary voltages into one analog voltage. Then it would be able to control the real world by generating a binary group of voltages, which get turned into an analog voltage by a D-to-A converter, which gets turned into a real world effector by a transducer. (See part (b) of Figure B-1, below.)

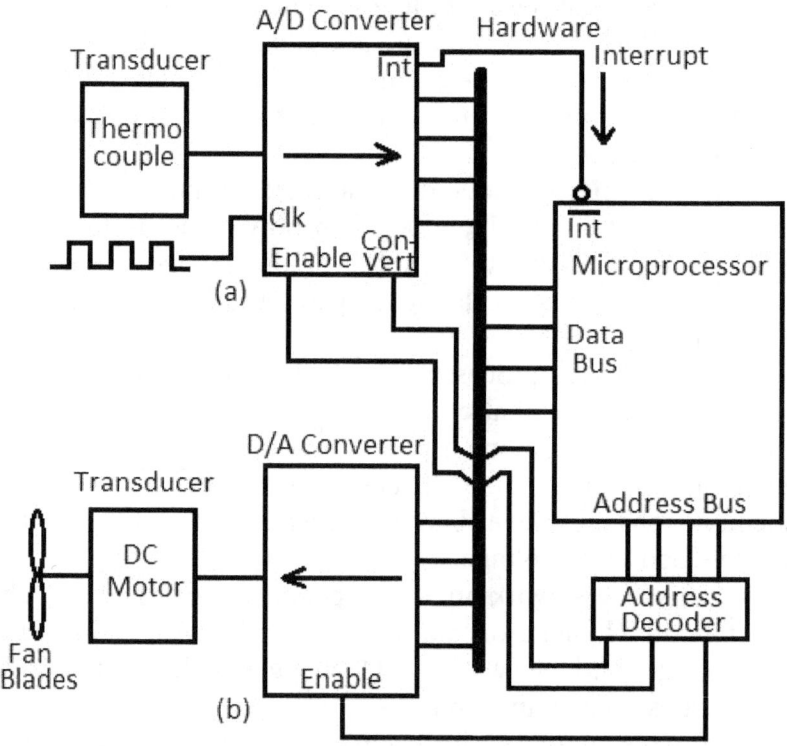

FIGURE B-1

Figure B-1 shows how one such system would work. Hopefully,

you are pleasantly surprised by how much of this system you already understand. It should greatly resemble Figure 10-4, which we studied in depth. The microprocessor uses the address bus and address decoder to enable the A/D converter, and read its input from the data bus. Also, the microprocessor enables the D/A converter, and writes information to it via the data bus. Both the A/D and D/A converters have our familiar D-type latches built into them.

The purpose of this circuit is for the microprocessor to read the temperature in part (a), the upper circuit. Then, if the temperature is too high, the microprocessor turns on the cooling fan in part (b), the lower circuit. Follow the arrows to understand the direction of information flow. Don't be confused by the extra "convert" and ""interrupt" lines on the A/D converter. Usually, you must first command the A/D converter to do a conversion ("convert"). When the conversion is complete, the A/D converter makes the "interrupt" line true, so that the microprocessor knows to read the results of the A/D conversion. You already understand interrupts, too! This circuit should be a real eye-opener, and help you see why billions of small microprocessors are sold.

I'll show you a simple design for a D/A converter, later. The design involves an op-amp integrated circuit, so let's start with that. See the picture, below:

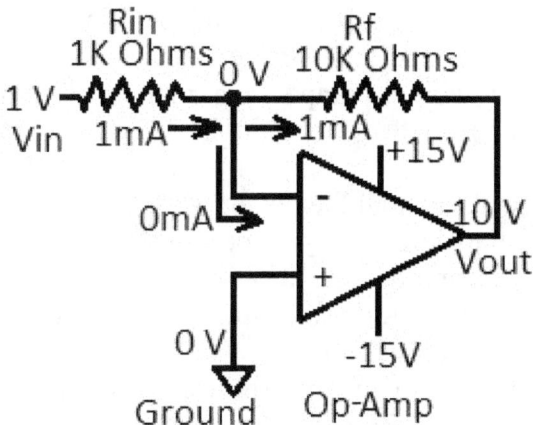

FIGURE B-2

The triangle circuit symbol in Figure B-2 is the op-amp. Don't confuse it with the symbol for a buffer or an inverter. Those two are binary devices. An op-amp is an analog device. The op-amp comes in an integrated-circuit body, with external pin connections to the outside world, so that different external components can be attached. This gives the op-amp the flexibility to become the core of many different kinds of circuits.

Op-amps brings great power and simplicity to analog circuit design. They usually simplify the mathematics needed to complete a design. They reduce the number of components needed for a circuit, in large part because they have many transistors inside their integrated-circuit body.

For example, the op-amp circuit configuration in Figure B-2 is an inverting voltage amplifier with a gain of 10. A voltage amplifier takes an input voltage and creates a bigger output voltage. In truth, an amplifier is a multiplier. In Figure B-2, the input voltage (Vin) of 1 volt is multiplied by 10, and output (Vout) as -10 volts. (Yes, *negative* 10 volts. We did say this is an *inverting* amplifier. We can create a non-inverting amplifier with a different op-amp design. Or, we can double-invert this circuit's output with a second inverting amplifier with a gain of -1.)

Look closely at Figure B-2 and you will see that the ratio of resistor Rf to resistor Rin is 10k ohms to 1k ohms. "k" is the metric symbol for "thousand." So, 10,000/1,000=10. Their ratio is 10, which, not coincidentally, is the gain of the circuit. Thus, one can learn to design amplifiers with op-amps in a matter of minutes. Can you change this design so that the gain is 20? (Answer, change Rf to 20,000 ohms.)

What's the secret behind the power, yet simplicity, of op-amps? I'll explain it now, but don't worry if you get lost. Just remember that, while the secret is complicated, the end result is simplicity. In the case of the inverting amplifier, for example, the gain is simply Rf divided by Rin, the ratio of two resistors.

The secret starts with an apparent absurdity. Internally, an op-amp is a differential amplifier, with a gain of about 100,000. In a differential amplifier, a subtraction takes place before the amplifier's multiplication. The voltage input at the pin labelled "+" minus the voltage input at the pin labeled "-" is multiplied by about 100,000, to produce an output voltage at Vout (the tip of the triangle on the right).

This is ridiculous. A difference in voltage of 1 volt from the "+" to the "-" input would produce an output voltage of 100,000 volts. That's impossible. The most this op-amp can produce from its Vout output is +15 volts or -15 volts (actually, a little less). That's because, as seen in Figure B-2, we are providing it with +15 volts and -15 volts as power supply inputs. In fact, for any difference in voltage of greater than .0001 volts between the "+" and the "-" inputs, this device is for the most part useless.

The second characteristic of op-amps is that the input resistance into the "+" and the "-" inputs is about 2 million ohms. Very little current will ever go into its inputs.

The secret to the amazing things an op-amp can do for us is hidden in a branch of engineering that most people have never heard of. It's called "control systems." The related job title is, "controls engineer." Control theory helps to prevent events like the collapse of the Tacoma Narrows Bridge, which broke into huge oscillations and fell down when the wind hit it just right. If you have never witnessed this, watch the video on YouTube.

Control theorists have developed a wonderful design trick called "negative feedback." A portion of the system's output is "fed back" and subtracted from the system's control input, to become the actual amplifier input.

This little trick has huge implications. It results in a system that is *self-correcting*. That means, if an external disturbance tries to change the desired output, the system reacts and compensates. The output is automatically restored to its desired state!

Negative feedback control systems are used to make furnaces that automatically run longer when it is colder; motors that turn with more torque when a conveyor belt has a heavier load; and car cruise control systems that maintain a constant vehicle speed for uphill or downhill roads.

In Figure B-2, trace the path from the output (Vout), going back through Rf, and into the input labeled "-". This is negative feedback. The output feeds back through Rf. Rf stands for "feedback resistor." It goes into the "-" input. This path is called the "closed loop".

The input, Vin, also goes to the "-" op-amp input. This junction is the point where the current from the feedback output current is subtracted from the input current, with the remainder going into the "-" input. Here's how the automatic error correction works. Assume that increased load current demand on the op-amp makes its output voltage decrease. That's a bad thing. We want output voltage to be reliably constant. This decreased output voltage causes less current to flow through Rf. Less Rf current, subtracted from Rin, leaves more current to go into the "-" input. More current entering "-" raises the output voltage. Output errors are thus self-corrected. A lowered output voltage gets automatically raised. That's what negative feedback control systems do.

Furthermore, the negative feedback also overcomes the problem of the internal gain of the op-amp being too high, in the range of 100,000. When we close the loop (insert the Rf path), the voltage and current difference between the "+" and "-" inputs is automatically kept small enough to keep the output voltage in a useable range. Remember, it's the difference between the voltage at "+" and "-" inputs that is multiplied to produce the output voltage.

Here's how all the tricks come together in the circuit of Figure B-2. The op-amp turns the node in the circuit at its "-" input into a "virtual ground."

What is a *virtual* ground? Well, the "+" input is attached to the *real* circuit ground. That's the point of reference for all our voltage measurements, often the negative of a battery or power supply. Due to negative feedback and the high internal gain of the op-amp, the op-amp circuit self-balances itself so that a voltage in the microvolt range is present at the "-" input. That's all it needs to be, since the differential gain is so huge (x100,000). Voltage-wise, the voltage at the "-" input is so close to 0 volts, it is "virtually" at ground potential.

However, even though the voltage at the "-" op-amp input is practically at ground voltage, it is not a direct path to true ground. Remember, almost no current comes into the "-" input. It has 2 million ohms of resistance.

The bottom line is: the junction where Rin, Rf, and "-" meet is kept at 0 volts, but virtually no current goes into the "-" op-amp input. Virtually all the current going through Rin winds up going through Rf.

Now, the left end of Rin has 1 volt applied to it. The right end of Rin is kept at virtual ground, 0 volts, by negative feedback control theory magic. So 1 volt is across Rin. Ohm's Law demands that the resulting current through Rin is $I_{in}=V_{in}/R_{in}=1/1000=.001$ amps.

We said virtually all the current going through Rin winds up going through Rf. That means .001 amps goes through Rf. Ohm's Law demands that the resulting voltage across Rf is $V_f=I_f x R_f=.001 x 10000=10$ volts. However, the conventional current flow is from left to right. Current moves to more negative energy levels. Since the left side of Rf is 0 volts, the right side has 10 volts less energy per coulomb of charge. So, Vout is -Vf, or -10 volts.

You may find all this baffling, but the result is very simple. We know that I_{in} effectively equals I_f, so we can call them both, "I." Basic algebra tells us we can substitute $I=V_{in}/R_{in}$ into $V_{out}=-IxR_f$, to get $V_{out}=(-V_{in}/R_{in})xR_f$. Re-arranging, we get $V_{out}=-(R_f/R_{in})xV_{in}$.

So, we have come full circle. When we began this discussion, I pointed out that the result would be this simple. The circuit gain is simply the ratio of Rf to Rin. In Figure B-2, the gain is -10. Hidden behind this simplicity is a load of complexity, as you have seen.

Now, let's forget about all this complexity, and focus on using the simplicity. Let's take the inverting amplifier from Figure B-2 and design from it a D/A converter. See the design in the figure below:

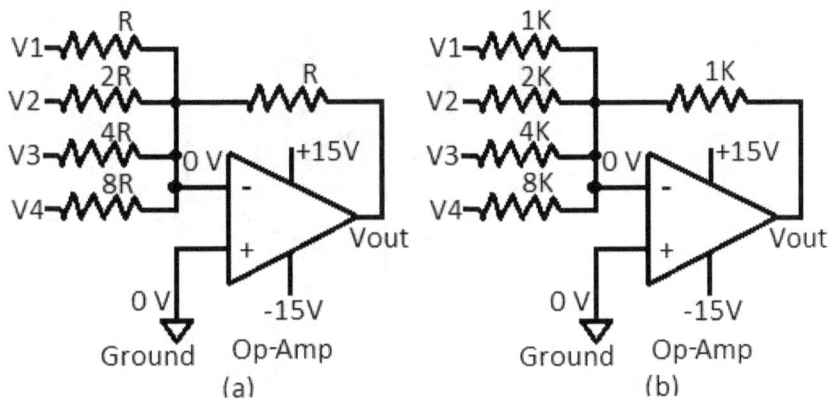

FIGURE B-3

Figure B-3, part (a), shows the D/A converter designed more abstractly. From it, you can see the mathematical relationship of resistor values. For whatever value of resistor value, R, that you choose, 2R must be a resistor value that is twice as big as R, 4R must be a resistor value 4 times the size of R, etc. Based on that relationship, I have designed an actual D/A converter in part (b), using resistor values 1k ohms, 2k ohms, 4k ohms, and 8k ohms. "k" is short for "kilo". "kilo" is a metric multiplier for 1 thousand. So, 8k ohms is

8,000 ohms.

The op-amp in Figure B-3 is a weighted, summing (inverting) amplifier. If all the resistor values in Figure B-3 were the same, it would be a regular summing (inverting) amplifier. The equation would be: Vout=(V1+V2+V3+V4).

Once again, we witness the simplicity of op-amp design. Want to add 10 voltages together? Get 11 resistors of equal value and build an op-amp summing amplifier like the one above. A discrete op-amp is smaller than a dime, and costs a quarter. Resistors cost a penny. No math is involved. If you don't like the negative sign in the output voltage, invert the voltage back to positive (double inversion: negative times negative equals positive) by attaching the following circuit to the output of Figure B-3:

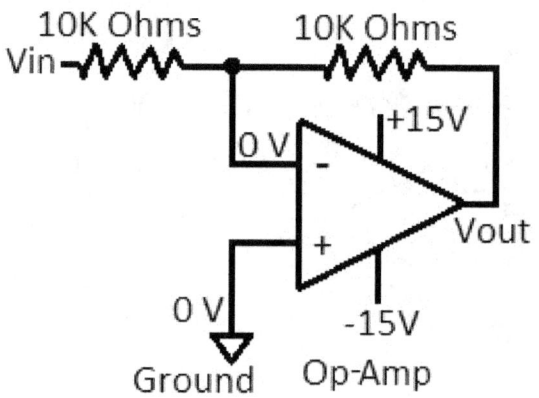

FIGURE B-4

Don't worry about paying more than a quarter if you add this circuit. The op-amp I found for 25 cents contains 2 op-amps in one 8-pin IC package.

However, the resistors in Figure B-3 are not all the same value. They are "weighted," in the same way that multi-bit binary numbers are weighted. In a binary number like 1011, as you move from the right-most bit to the left-most bit, the value of each bit position doubles. Representing a 4-bit binary number as ABCD, the D bit is worth decimal 1, the C bit is worth 2, the B bit is worth 4, and the A bit is worth 8. The resistors in the op-amp circuit in Figure B-3 follow the same doubling sequence. That should make sense, because we are trying to turn a 4-bit binary number into an analog voltage.

The op-amp circuit in Figure B-3, part (b), is easy to understand. We can analyze it as if it were four separate op-amp circuits like the one in Figure B-2. In other words, we can cover up the 2k-, 4k-, and 8k-ohm resistors, and find the gain for the V1 voltage input to be -1k/1k =-1. Then, we can cover up the 1k-, 4k-, and 8k-ohm resistors, and find the gain for the V2 voltage input to be -1k/2k =-.5. We can cover up the 1k-, 2k-, and 8k-ohm resistors, and find the gain for the V3 voltage input to be -1k/4k =-.25. We can cover up the 1k-, 2k-, and 4k-ohm resistors, and find the gain for the V4 voltage input to be -1k/8k =-.125.

Since the V1, V2, V3, and V4 inputs are all digital, they will be either 0 volts or 5 volts. If they are 5 volts, the V1 input will deliver -1x5=-5 volt at Vout; the V2 input will deliver -.5x5=-2.5 volts at Vout; the V3 input will deliver -.25x5=-1.25 volts at Vout; and the V4 input will deliver -.125x5=-.625 volts at Vout.

Alternatively, if we do the same analysis on Figure B-3, part (a), we come up with equations for each separate input's gain: for the V1 input, Vout=-R/R=-1xV1; for the V2 input, Vout=-R/2R=-.5xV2; for the V3 input, Vout=-R/4R=-.25xV3; and for the V4 input, Vout=-R/8R=-.125xV4.

The easy part is that we can simply add these four results together to get the actual behavior--and the equation--of this circuit. Each of the four circuits contributes its own current through the feedback resistor. These currents simply add together, then get multiplied by R to create Vout.

So, adding together the four results from two paragraphs back, we produce this equation for this weighted, summing amplifier:

Vout=-(V2+.5xV2+.25xV3+.125xV4), or
Vout=-(V1+V2/2+V3/4+V4/8).

We can see from the above results that the least significant bit input must be applied as V4, and the most significant bit input must be applied as V1.

The analysis wouldn't be this simple for most circuits designed to add voltages together. Any one input voltage would interfere with the other three input voltages. But, the op-amp's virtual ground at the "-" input prevents this interference. (Application of the superposition theorem, which I won't cover, coupled with an understanding of the virtual ground, proves this.)

Let's make a table from the above results. We can create the table from the numerical results above. Or, we can create it from the above equation, remembering that V1, V2, V3, and V4 can only be 0 volts or 5 volts.

V1	V2	V3	V4	Vout	Vout x-1.6
0	0	0	0	0	0
0	0	0	5	-.625	1
0	0	5	0	-1.25	2
0	0	5	5	-1.87	3
0	5	0	0	-2.5	4
0	5	0	5	-3.12	5
0	5	5	0	-3.75	6
0	5	5	5	-4.37	7
5	0	0	0	-5	8
5	0	0	5	-5.62	9
5	0	5	0	-6.26	10
5	0	5	5	-6.87	11
5	5	0	0	-7.5	12
5	5	0	5	-8.12	
5	5	5	0	-8.75	
5	5	5	5	-9.37	

FIGURE B-5

The table in Figure B-5 clearly demonstrates the D/A conversion of the circuit in Figure B-3. The voltages in columns labeled V1 through V4 are all the possible binary input voltages, in ascending binary numerical order. The circuit output voltage is displayed in the column labeled Vout. This analog output voltage steps up (actually down, since it is negative-going) by .625 volts, for each binary count increment.

We often don't care about the negative sign of the output voltage. And, we usually don't care about the fact that the analog output voltage is not the exact numerical value corresponding to the binary input value. We usually just care that the analog Vout's increases are proportional to the digital Vin's increases.

However, if the negative polarity and relative proportionality of input to output are a problem, you can add a second op-amp circuit to the output of the first:

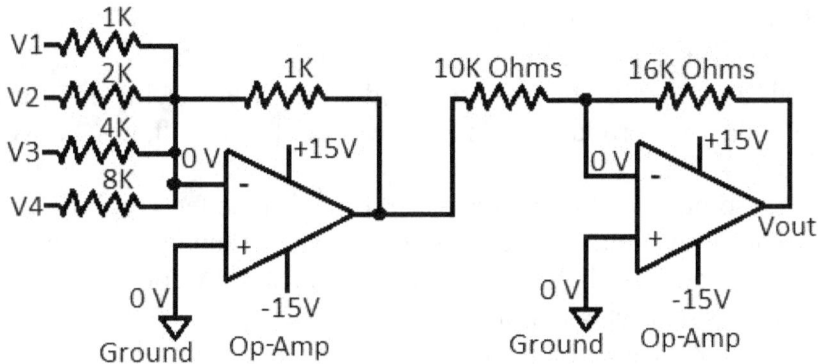

FIGURE B-6

I have described this solution, before. But, this time, the gain of the second op-amp, on the right, is times negative 1.6 (x-1.6). This output inverts the negative values of the first op-amp back to positive. Its gain, x-1.6, creates an output voltage that is the decimal analog equivalent of the binary input voltage.

The last column in the table in Figure B-5 displays the analog output voltage, Vout, from Figure B-6. For example, input voltages of V1=5v, V2=0v, V3=0v, and V4=5v correspond to an input of 1001, in binary logic. Binary 1001 equals decimal 9. Nine volts comes out of Vout in Figure B-6, and is listed in the table of Figure B-5.

I'll mention a few drawbacks in this D/A converter design. First, you'll notice that, in Figure B-5, I don't list output voltages above 12 volts. The circuit in Figure B-6 can't reach 15 volts. Maximum op-amp output voltages are less than their power supply voltages. My design, above, used +15 volts and -15 volts power supplies. Specific op-amp specification sheets would tell a designer how close to 15 volts the output can reach. One solution to this problem is to use +18 volts and -18 volts power supplies. Then, our output voltages could reach 15 volts.

Second, the *resolution* of the above design is not very good. D/A resolution is one divided by the maximum binary count. In our example, the maximum binary count is 1111 (decimal 15). So, the resolution is 1/15=.066667. So, if our maximum output voltage is 15 volts, the D/A converter's output steps are all 1 volt apart. That's because 15 times .066667 equals 1 volt. From this circuit, we can't produce output voltages of 3.3 volts, or 3.7 volts. The closest we can get is the nearest resolution values of 3 volts or 4 volts.

This would not sound very good if we were trying to use D/A conversion to reproduce sounds. We need those voltages between 3 volts and 4 volts.

The solution is to create a circuit with better resolution. That means our D/A circuit must accommodate more binary bits. The above circuit was good to help you understand how D/A conversion works, because it is relatively simple.

However, it does not scale well. To accommodate more bits with this design, we must keep doubling the size of the resistors. But, if we want to create a 12-bit wide A/D converter (resolution=.00024) from this circuit, some of the resistor values would be too big.

Op-amps don't work well with resistor values that are too big. The magic properties of the op-amp become negated, and the accuracy gets destroyed. Op-amps also don't work well with resistor values that are small. Those resistors draw too much power, and get hot.

D/A conversion based upon the R/2R design would be better to accommodate improved resolution. That design also builds around an op-amp, but uses resistors that don't spread over a wide range. It's just harder to explain, requiring from the student a knowledge of Thevenin's theorem. A totally different D/A conversion design is one based upon filtered pulse-width modulation. I won't discuss these, here.

Third, realize that our op-amp design requires accurate resistors. *Precision* is the term we use to describe how close out component is to its listed value.

Also, the 5 volts and 0 volts input values need to be as accurate as possible. This is not as simple as it sounds. Digital specifications that define a valid high and a valid low voltage can be sloppy. For better accuracy, voltage regulator circuits can be added to keep voltages close to 5 volts.

Please don't get too hung up on the three drawbacks. The solutions I pointed out are simple to implement. Don't miss the point of this appendix: digital-to-analog conversion greatly extends the power of microprocessors. With it, there's almost no end to the number of things the microprocessor can control: robots, cars, assembly lines, 3D printers. The list goes on and on.

APPENDIX C
A/D CONVERSION

In this bonus circuit I will describe how analog-to-digital conversion works. But, in order to do that, I will first explain comparators. A comparator is an integrated circuit that compares two analog voltages, Va and Vb. If Va is greater than Vb, the comparator's output becomes a high voltage. If Va is less than Vb, the comparator's output becomes a low voltage. Thus, a comparator compares analog input voltages, but produces digital (binary, 2-state) output voltages. Refer to the picture, below:

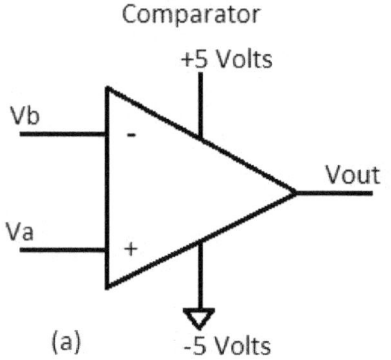

Va	Vb	Vout
+3v	+2v	+5v
+2v	+3v	-5v
-3v	-2v	-5v
-2v	-3v	+5v

(a)   -5 Volts   (b)

FIGURE C-1

Looking at Figure C-1, part (a), we find the circuit diagram for a comparator. Notice that the Va analog voltage input enters the "+" input, and the Vb analog input voltage enters the "-" input. Notice also that two power supply voltages are provided: +5 volts and -5 volts.

We can describe the comparator's behavior in terms of signed mathematical subtraction. We say that the comparator subtracts Va minus Vb. If the result is a positive number, the output voltage becomes the same as the +5 volts power supply. If the result is a negative number, the output voltage becomes the same as the -5 volts power supply.

The table in Figure C-1, part (b), gives a few examples to help you see this more clearly. Remember from math class that -2 is greater than -3. Also remember that -(-2)=+2. In row one, +3-(+2)=+1, which is positive, so the output goes to +5 volts. In row two, +2-(+3)=-1, which is negative, so the output goes to -5 volts. In row three, -3-(-2)=-1, which is negative, so the output goes to -5 volts. In row three, -2-(-3)=+1, which is positive, so the output goes to 5 volts.

The comparator builds upon something you already know. For, the comparator is an op-amp without the negative feedback connection! That being said, when shopping for a comparator, don't buy an op-amp. Buy a component labeled, "comparator." The comparator has been optimized to switch from a high output voltage to a low output voltage (or vice-versa) more quickly than an op-amp used without feedback.

If you remember, we said that an op-amp is a difference amplifier. It subtracts the "-" input from the "+" input. Then it amplifies (multiplies) that difference by a ridiculously high internal gain of times 100,000. Thus, even a difference of greater than 50 microvolts (.000050 volts) would cause the output voltage of the above circuit to reach its +5 volts or -5 volts limitation.

Well, the comparator takes advantage of that problem. It doesn't "cure" it, by adding negative feedback. If the voltage on the "-" input is just a smidgen more than the voltage on the "+" input, the output voltage goes to as low as it can, which is in this case is the -5 volts limitation of its negative power supply. If the voltage on the "-" input is just a smidgen less than the voltage on the "+" input, the output voltage goes to as high as it can, which is in this case is the +5 volts limitation of its positive power supply. (Remember from previous discussions that, due to internal losses, outputs never quite reach all the way to +5 volts or -5 volts.)

Now, the two power supply voltages on the comparator don't have to be +5 volts and -5 volts. They can be, for example, +15 volts and -15 volts, like we used for our previous op-amp circuits. Some comparators are optimized to work at power supplies of +5 volts and 0 volts.

A comparator that only has two output voltages--+5 volts and 0 volts--interfaces perfectly with digital electronics circuits. Such a component makes a perfect connection between the world of analog on its inputs and the world of digital on its output. We will use such a device in our design of an A/D converter.

The picture below shows one design for an A/D converter:

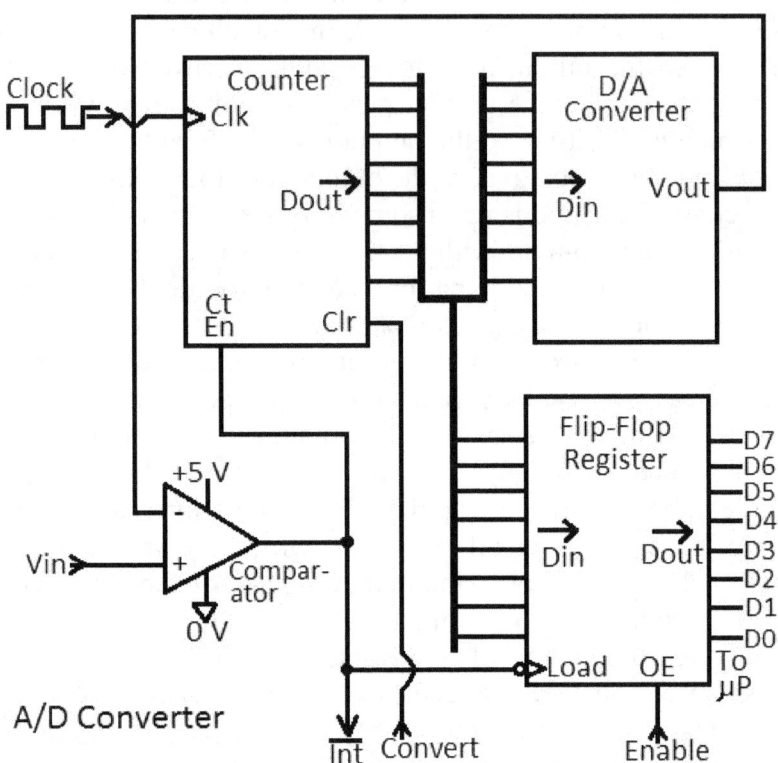

FIGURE C-2

This is one design, chosen from several, of an A/D converter. I have chosen this one for the same reason I have chosen many of the previous designs in this book: it builds upon sub-circuits you already know. One goal from the start has been for you to experience how digital electronics and microprocessor circuits build gradually from simplicity to complexity. Each circuit is built from building blocks of simpler circuits. Building blocks contain simpler building blocks, which contain simpler building blocks, which contain simpler blocks, etc.

When you look at Figure C-2, you find subsystems that you already know. See Figure 16-4 to review counters; Figure B-6 to review D/A converters; and Figure C-1 to review comparators. See Figure 7-13 to review D latch registers. Here we use a register made of flip-flops instead of latches. But, you know flip-flops from Figures 15-1 and 15-2. Our register has tri-state outputs, to take part in a microprocessor bus system. You know tri-state outputs from Figures 7-19 and 7-20. Finally, refer to Figure B-1 to see how this A/D converter communicates with a microprocessor-based system, through input and output lines Int, Convert, Enable, as well as the 8 data output lines from the flip-flop register.

As an overview, consider that the purpose of an analog-to-digital converter is to change an analog voltage into a multi-bit digital voltage. The analog voltage enters the comparator as Vin, on the lower left side of Figure C-2. The digital voltage comes out the flip-flop register as D7 through D0, on the lower right side of Figure C-2.

Notice that the power supplies on the comparator are +5 volts and 0 volts. As I described earlier, this limits the comparator output to +5 volts or 0 volts, perfect for the digital logic circuits it controls. For simplicity, we will assume Vin's analog range is 0 volts through +5 volts, too. Just be aware that there are other options available that allow for a range of inputs from negative through positive voltages, to accommodate sine wave input voltages. (One solution: We can shift input voltages up, out of the negative, with an op-amp summing circuit.)

Next, look at Figure B-1 to see how the microprocessor connects to, communicates with, and controls the A/D converter. Focus on the Convert, Enable, and Int lines. Here's what the microprocessor does:

1.) Pulses the Convert line high (then back low), to clear the counter to 00000000 and start the conversion process.
2.) Internally enables the A/D conversion hardware interrupt recognition circuitry, so the microprocessor can respond to the A/D converter's interrupt request (Int).
3.) Does other stuff, while awaiting the A/D converter's interrupt.
4.) Responds to the A/D converter's interrupt by *first* disabling the A/D conversion hardware interrupt recognition circuitry, so that the microprocessor temporarily ignores further A/D interrupt requests.
5.) And *second* by raising the Enable line high, to read the D7 through D0 lines out of the 8-bit flip-flop register.

To reiterate, the D7 through D0 lines out of the flip-flop register are the digital equivalent of the analog input voltage, Vin.

Step 1, above, clears the counter to an output of 00000000. The D/A converter receives this binary input and converts it to an analog output voltage of 0 volts. Be careful to avoid confusion, here. Yes, this is a D/A converter, buried in the heart of the A/D converter. Confusion can result if your brain doesn't keep track of my references to A/D versus D/A.

The 0 volts out of the D/A converter feeds into the "-" input of the comparator. Since 0 volts is less than the positive voltage on the comparator's "+" input, the comparator's output becomes +5 volts. That +5 volts allows the counter to count by raising its "Ct En" input high. That same +5 volts from the comparator also turns off the interrupt request to the microprocessor since our design requires a low (0 volts) signal to request an interrupt on the Int line. (Signified by the bar over the Int label and the bubble on the microprocessor input.)

The counter counts the external clock pulses, perhaps the same clock that runs the microprocessor. As the binary count rises, the voltage out of the D/A converter also rises. It keeps rising on the comparator's "-" input, as it strives to reach the input voltage on the comparator's "+" input.

When the voltage on the comparator's "-" input finally exceeds the voltage on the comparator's "+" input by just one step up out of the D/A converter, everything changes. The output of the comparator goes to 0 volts. That 0 volts stops the counter from counting, while maintaining the last binary count value.

That count value is precious to us. It is the binary representation of the voltage we are digitizing, the input voltage Vin. How do we know this? Because it is the binary value that caused the D/A converter to produce a Vout output voltage equal to the A/D converter's Vin input voltage. It must be so, because the comparator announced it, when its output went to 0 volts.

The comparator's output going to 0 volts does more. Its falling edge stores our precious count from the counter into the flip-flop register. It's stored there until the microprocessor is ready to read it from the data bus.

The comparator's output going to 0 volts does a third thing. It is also the NOT Int signal that, when low, requests an interrupt from the microprocessor. Remember, the microprocessor has been doing other chores, while waiting for the NOT Int line to go low.

Finally, the microprocessor responds to the interrupt by running interrupt processing code that reads the tri-state outputs of the flip-flop register. It does this by generating a high logic level voltage on Enable, activating the flip-flop register's OE (output enable) line.

The microprocessor now has the results of the A/D conversion in its accumulator. The NOT Int line is still active (Low), but that's OK. Per step 4, above, the microprocessor is ignoring it. The microprocessor won't pay attention to it again until after it starts the next conversion, when it pulls the Convert line high, which deactivates (raises high) the NOT Int line and makes the interrupt request go away.

What does this A/D conversion value, now in the microprocessor's accumulator, look like? Let's figure it out for ourselves. Let's assume that the A/D converter's internal D/A converter works on an output range of 0 to 5 volts. From this we determine that the resolution of the 8-bit D/A converter is 1/255th of 5 volts, or .0196 volts.

From this information, we can calculate the expected value that the microprocessor's accumulator holds. Let's say the analog input voltage from the thermocouple is +1 volt. After A/D conversion, what value should the microprocessor have read?

Well, with a resolution of .0196 volts, the least significant bit on the D/A's input is worth .0196 volts. The second-least most significant bit is worth twice .0196, the third most significant bit is worth four times .0196, etc. From this information, we can make the first two rows of the following chart:

x128	x64	x32	x16	x8	x4	x2	x1
2.5088	1.2544	.6272	.3136	.1568	.0784	.0392	.0196
0	0	1	1	0	1	0	0

FIGURE C-3

The bottom row of the chart is our answer. +1 volt from the

thermocouple is read as 00110100 from the A/D converter. To arrive at that answer, I simply used the trial-and-error technique we used in chapter nine to solve decimal-to-binary conversion problems. Proceeding from left to right, I put a 1 under the first number that is less than the number we are converting to binary, which in this case is +1 volt. .6272 is that number. Next, still proceeding from left to right, I put a 1 under any other numbers which, *when added to our running tally of numbers chosen from row 2*, just exceeds +1 volt.

Using this technique, I found that .6272+.3136+.0784=1.0192. That's the first value higher than +1 volt. With 1's under those three values, and 0's under the others, 00110100 is the first count that causes the comparator to trip when analog +1 volt enters the A/D converter for conversion. In decimal, 00110100 is equivalent to 52. When we multiply 52 times .0196 we get 1.0192, which proves our solution to be correct.

So, the microprocessor reads +1 volt as 00110100. What does the programmer do with this? In our design shown in Figure B-1, we see that the +1 volts comes from a transducer called a thermocouple. The programmer must consult the specifications on the thermocouple to determine how voltages correspond to temperatures. Specific voltage ranges correspond to specific temperature ranges. The programmer's code can test the binary value from the A/D converter to determine which range of voltages, and hence which range of temperatures, the binary value falls in. For higher temperature ranges, the programmer's code can send proportionally higher values to the D/A converter that controls the cooling fan, making it spin faster.

I hope you enjoyed exploring how this circuit works. There is so much to it, and yet an understanding of it builds from things we have previously covered. You can think it through, work backwards, even drill down to the level of transistors and electrons to enjoy the complete picture of how this circuit works, in all its glorious detail.